P9-AEU-318

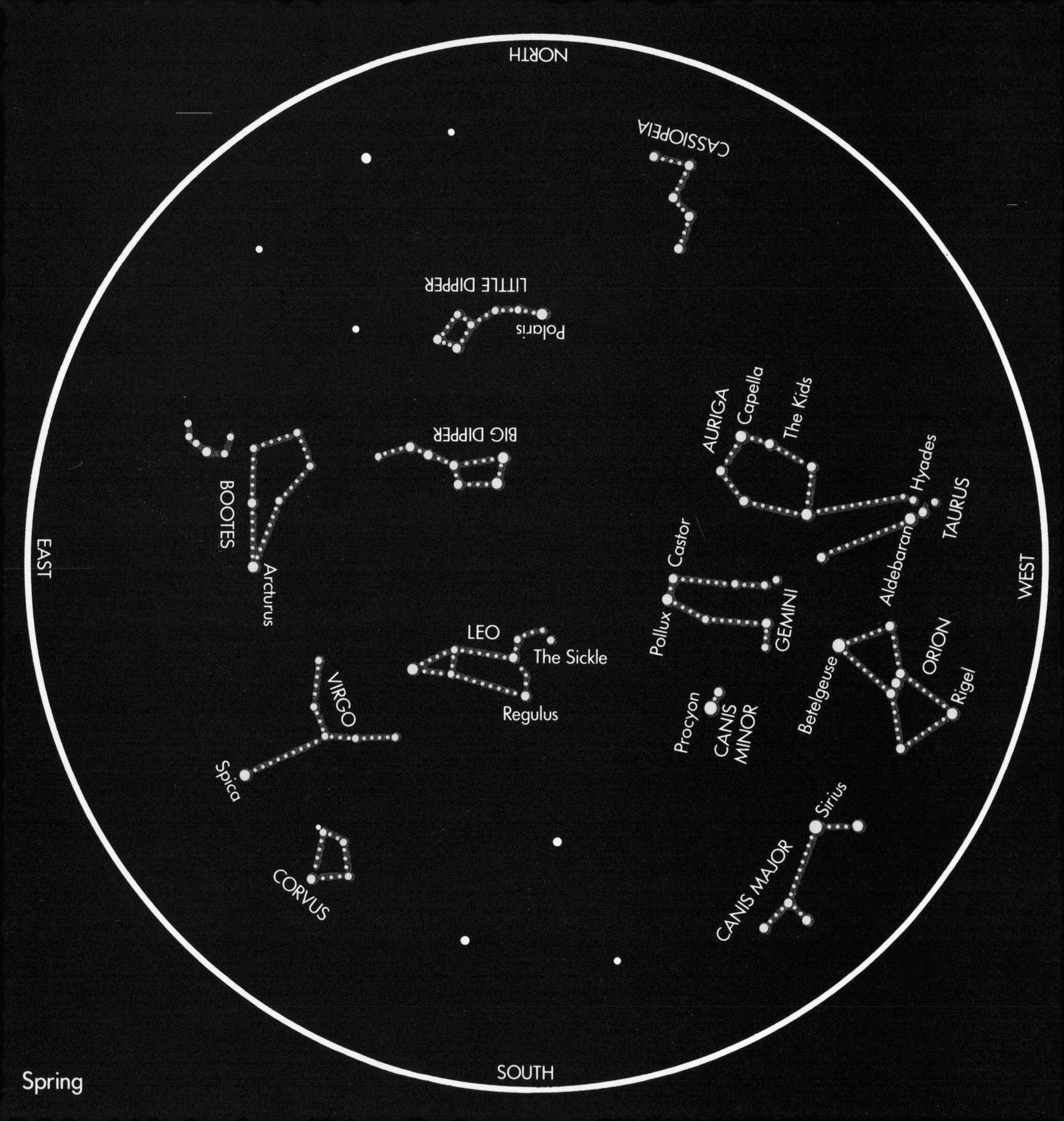
NORTH
CASSIOPEIA
LITTLE DIPPER
Polaris
BIG DIPPER
BOOTES
Arcturus
EAST
AURIGA
Capella
The Kids
Hyades
TAURUS
Aldebaran
Castor
Pollux
GEMINI
WEST
ORION
Betelgeuse
Rigel
LEO
The Sickle
Regulus
VIRGO
Spica
Procyon
CANIS MINOR
Sirius
CANIS MAJOR
CORVUS
SOUTH

Spring

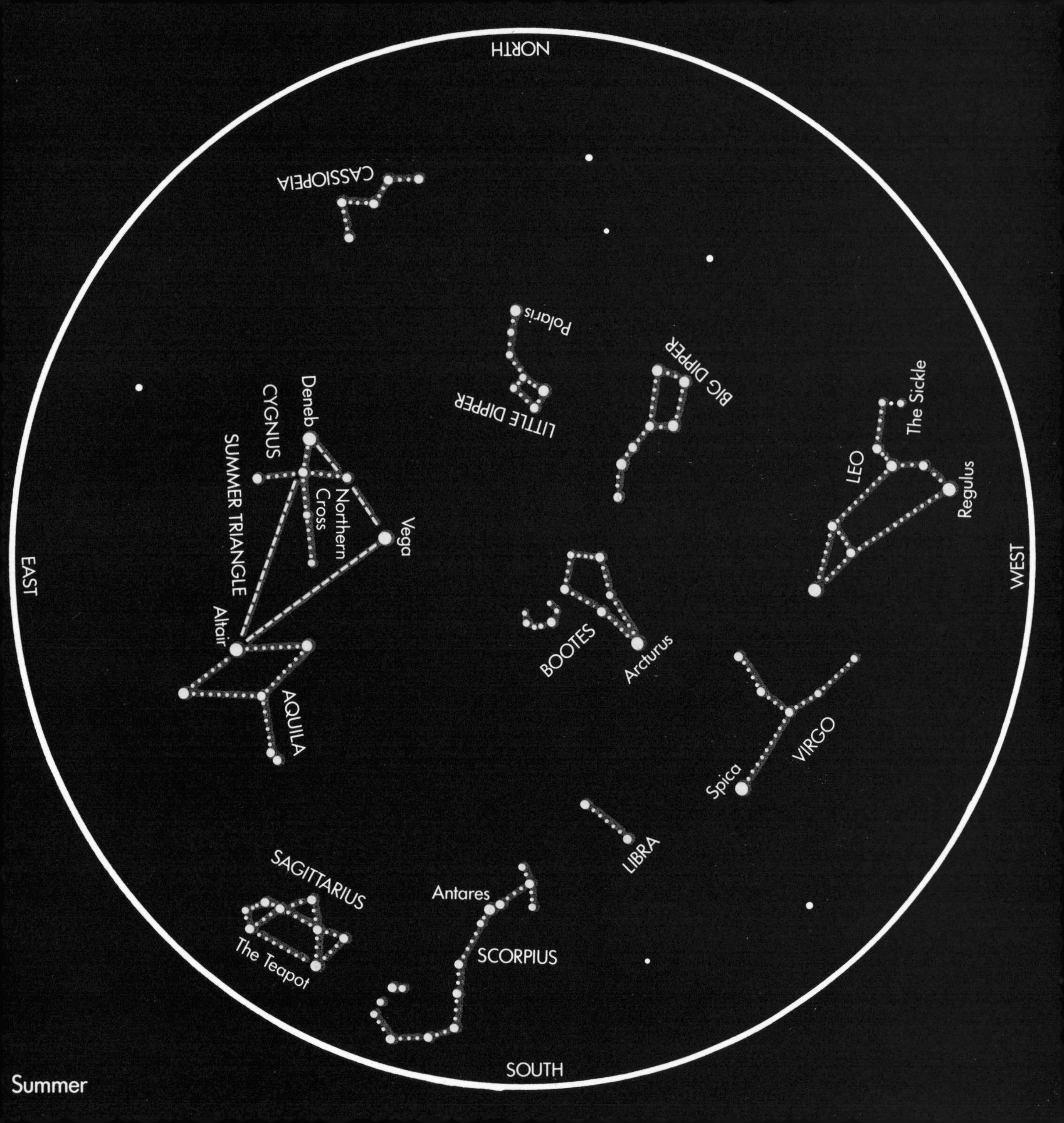
NORTH
CASSIOPEIA
Polaris
LITTLE DIPPER
BIG DIPPER
The Sickle
LEO
Regulus
Deneb
CYGNUS
SUMMER TRIANGLE
Northern Cross
Vega
EAST
WEST
Altair
BOOTES
Arcturus
AQUILA
VIRGO
Spica
LIBRA
SAGITTARIUS
The Teapot
Antares
SCORPIUS
SOUTH
Summer

Discover

PLANETWATCH

A YEAR-ROUND VIEWING GUIDE TO THE NIGHT SKY
WITH A MAKE-YOUR-OWN PLANETFINDER

Discover

PLANETWATCH

A YEAR-ROUND VIEWING GUIDE TO THE NIGHT SKY WITH A MAKE-YOUR-OWN PLANETFINDER

Clint Hatchett

Illustrations by Brian Sullivan

NEW YORK

A RUNNING HEADS BOOK

DISCOVER PLANETWATCH
was conceived and produced by
Running Heads Incorporated
55 West 21 Street
New York, New York 10010

Creative Director: Linda Winters
Senior Editor: Thomas G. Fiffer
Designer: Jack Tom
Managing Editor: Jill Hamilton
Production Associate: Belinda Hellinger

Library of Congress Cataloging-in-Publication Data

Hatchett, Clint.
Discover planetwatch : a year-round viewing guide to the night sky with a make-your-own planetfinder / Clinton W. Hatchett : illustrations by Brian Sullivan.—1st ed.
p. cm.
"A Running heads book"—T.p. verso.
Includes bibliographical references.
ISBN 1-56282-874-6
1. Astronomy—Amateurs' manuals. 2. Astronomy—Observers' manuals. I. Sullivan, Brian. II. Title.
QB63.H35 1993 92-29623
523.2—dc20 CIP

Typeset by Trufont Typographers Inc.
Color separations by Vimnice Printing Press Co., Ltd.
Printed and bound in Hong Kong

FIRST EDITION
10 9 8 7 6 5 4 3 2 1

To Grace, my FIRST love, and our five "Hatchlings" who make life complete.

AUTHOR'S ACKNOWLEDGMENTS

Special thanks to Brian Sullivan for going above and beyond to create illustrations for a "non-artistic" writer, and to Lynn Rice for making my "real" job a little easier so I could find bits of time for this project. Thanks, too, to Tom Fiffer for his patience and encouragement, and to everyone else at Running Heads.

ILLUSTRATOR'S ACKNOWLEDGMENTS

Special thanks to Suzanne Chippindale, astronomer/producer, Hayden Planetarium, New York; Dennis Mammana, production manager, Reuben H. Fleet Space Theater and Science Center, San Diego, California; and Victor Costanzo, production designer, Strasenburgh Planetarium, Rochester, New York.

CONTENTS

For hundreds of years after the dawn of modern astronomy, watching the sky remained the province of serious scientists. The knowledge and equipment needed to make accurate and useful observations of planets and stars were unavailable to the average person. But, today, with simple optical instruments and devices made from ordinary household items, anyone can indulge an interest in the sky and enjoy hours of fun—and practical—planetwatching.

Discover Planetwatch is a hands-on, project-oriented guide to understanding what goes on above us. The book is divided into sections—Moon, Sun, Stars, and Eclipses; The Solar System; The Inner Planets; and The Outer Planets—from which you can choose projects of particular interest. Projects are offered at three different levels—basic, intermediate, and advanced—to provide a variety of activities for both beginning and experienced astronomers. Even the advanced projects are easy, and each project includes a list of all the equipment you will need to carry it out. To complete many of the basic projects, you will not need a telescope, or even binoculars. Most of the astronomical data you will need for the projects is provided in maps, charts, and tables within the book. For information not given, references—and places to find them—are suggested in the Bibliography (page 122). The Tools of the Trade section (page 12) tells you how to construct or where to find the equipment for the projects, and the Glossary (page 124) gives helpful definitions of scientific terms.

To use and enjoy *Discover Planetwatch* to its fullest, you should begin by becoming familiar with the geography of the sky and learning a few basic guidelines for astronomical observation. Then choose an aspect of the sky you would like to explore and find a project that piques your interest.

Astronomical Observation Guidelines

One of the crucial components of any scientific observation is constructing a record that states clearly what you saw, the conditions under which you saw it, and exactly when you saw it. The Observation Form (opposite) provides space for all the essential information you should record each time you make an astronomical observation. If this form looks daunting, don't worry. Most of the positional information you record will derive from a simple instrument, the sighting stick, that you can construct from instructions given in the Tools of the Trade section, and the scientific terms are defined in the Glossary (page 124). Make some copies of this form, or prepare your own version, before you begin the projects. To obtain exact times and timings, use a digital watch, or for greater accuracy, time signals available over shortwave radio.

You never know what you may see in the sky, and for your observations to be useful, perhaps even to professional astronomers, you must record them properly. Most comets and many asteroids are discovered by amateur astronomers making careful observations of the sky and recording what they see.

OBSERVATION FORM

Date ______________________ Location ______________________ Equipment ______________________

Start of session ______________________ Weather conditions ______________________ Main object ______________________

Sighting angles:

Object: ________	Object: ________	Object: ________	Object: ________
Time: ________	Time: ________	Time: ________	Time: ________
Altitude: ________	Altitude: ________	Altitude: ________	Altitude: ________
Azimuth: ________	Azimuth: ________	Azimuth: ________	Azimuth: ________

Timings (for occultations, transits, eclipses):

Began observing: ______________________
1st Contact: ______________________
2nd Contact: ______________________
Maximum: ______________________
3rd Contact: ______________________
4th Contact: ______________________
End of observation: ______________________
Special observations: ______________________

Drawings (use circles for entire object or field of view; mark position in the sky or identify known objects in field of view if possible):

Object: ________	Object: ________	Object: ________
Note #: ________	Note #: ________	Note #: ________
Start: ________	Start: ________	Start: ________
End: ________	End: ________	End: ________

Notes:

Observing condition details:

The Geography of the Sky

To observe the sky successfully, you must understand its basic geography.

As time passes, whether hours or years, planets and stars seem to move, and the picture of the sky changes. To illustrate the apparent movement of celestial bodies (objects in the sky) and to explain why it happens, astronomers have created an imaginary sphere, known as the *celestial sphere*, on which to draw lines that serve as a map of the sky. At the celestial sphere's center is you, the observer, on Earth. The dome of the sky forms the top half of the sphere above you. Cutting through the sphere, dividing the sky exactly in half, is the horizon. Depending on your location on Earth, the time of night, and the season of the year, a particular portion of the sky will be visible. This way of looking at the sky helps us understand star maps, which incorporate *great circles* that divide the sky in various ways and locate important places in it. The most important of these circles are shown and labelled in Figure 1. On pages 20–23 you will find four simplified star maps, one for each season, indicating the major constellations that will be visible, along with instructions on how to orient the maps to the sky.

The Celestial Sphere, an imaginary globe that extends Earth's latitude, equator, and the horizon into space, gives observers a reference grid for finding objects in the sky.

The celestial sphere shows us how astronomers work with the sky. For the purposes of mapping and locating objects, the sphere, like the Earth, is divided into segments. These segments are defined by what appear to be extensions of the Earth's latitude lines drawn out into space. These lines, known as *parallels of decli-*

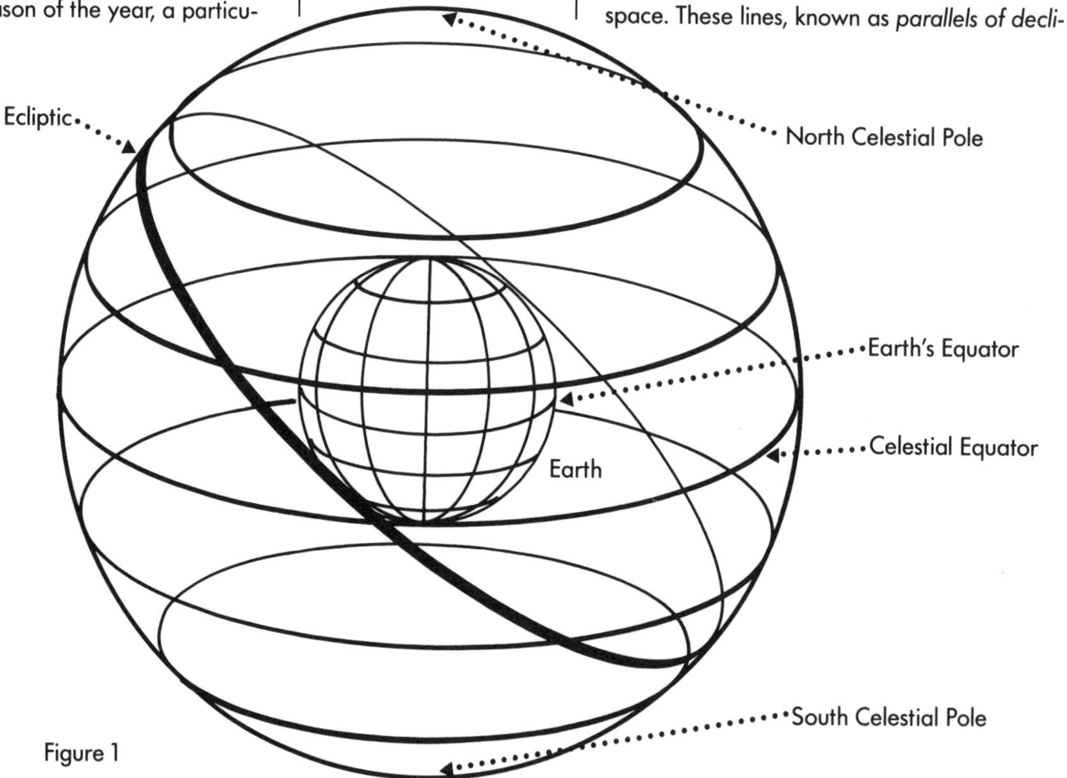

Figure 1

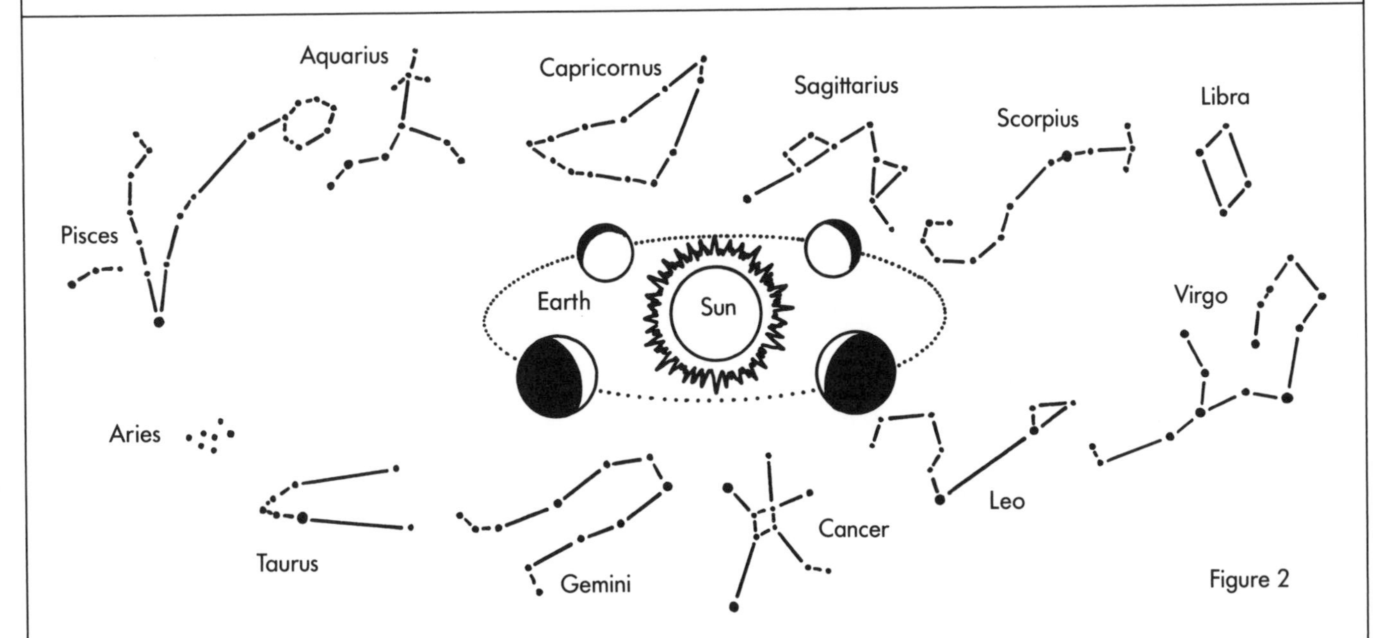

Figure 2

nation, actually do line up perfectly with the latitude markings on our globe, and they serve the same purpose. For example, the celestial equator lies exactly above the Earth's equator. Hence, if you were standing on the Earth's equator, the celestial equator would be an arc starting at the east point on the horizon, traveling straight overhead to the zenith (the point directly overhead), then descending to the west point. No matter where you are on Earth, the celestial equator, like all the other great circles, is an arc that curves across the sky.

Trying to find the North Star, Polaris, provides a good example of how the celestial sphere works. On the sphere, Polaris is very near the point in the sky directly above the North Pole on Earth. If you are at a latitude of 45 degrees north, the North Star will be 45 degrees above the north point on your horizon. If you are near the equator or farther south, you will never see Polaris in your sky.

As Earth circles the Sun, the Sun seems to move against the stars that lie along the ecliptic, which include the twelve constellations of the zodiac.

Because we will be observing our nearest neighbors in space, the Sun, the Moon, and the planets, the most important of the great circles to find and become familiar with is the *ecliptic*. This is the path that the Sun seems to travel in the sky against the familiar twelve constellations of the zodiac. Visualize the ecliptic as a line out in space (around the outside of the celestial sphere) that encircles the Earth. The sun appears to make a complete circuit year of the celestial sphere once a year, moving from one constellation to the next along the ecliptic. In truth, the Sun doesn't move along the ecliptic; the movement we see is actually the result of the Earth orbiting the Sun. Figure 2 shows why the Sun seems to move in front of certain stars as the year passes. See if you can find some of the constellations that lie along the ecliptic for the current season of the year.

With just this basic knowledge of the sky, you are ready to begin your planetwatch!

Fig. 38
Fig. 16
Fig. 31
Fig. 6
Fig. 32
Fig. 33
Fig. 26
Fig. 27
Fig. 1
Fig. 24
Fig. 23
Fig. 36
Fig. 29
Fig. 28
Fig. 2
Fig. 4
Fig. 3

SECTION 1
TOOLS OF THE TRADE

TOOLS OF THE TRADE

Most of the projects in this book require only basic observing equipment. In fact, you will be able to make some of the things you need. The book provides an Observation Form (page 9), that you can photocopy, star maps (pages 20–23), and various tables and illustrations for use with the projects. Other equipment items necessary for most of the projects are paper, pencils (regular and colored), a ruler, and several reference books and periodicals. A list of vendors for telescopes, binoculars, eclipse filters or glasses, and computer programs, is given as well (page 120). Most of the books and both magazines mentioned can be found in major public libraries, and full information on them is given in the Bibliography (page 122).

An accurate time source is also vital to many of the projects. In most cases, you can use your own wristwatch, but you should synchronize it with a known accurate source (such as telephone time service) if possible. WWV broadcasts National Bureau of Standards time signals and provides the kind of accuracy needed to record occultations for scientific use, but you must have a weather radio to receive this station. If you have a shortwave radio, the signals can be received by tuning to 5, 10, or 15 megahertz.

Projects 6 and 7, although basic, require a few unusual items, such as a standard basketball, some large balloons, wooden beads, marbles or ball bearings, pins, index cards, yarn, scissors, a yardstick or tape measure, and transparent tape. These can be easily found at appropriate stores, but for the items used to represent planets, you should take the size information in Table 6 and do some measuring as you shop.

Sighting Stick

Measuring angles in the sky is an important aspect of many projects. The sighting stick is a simple device that you can build very inexpensively from material in the scrap pile of a woodshop or from pieces you may have around the house. You will need to use the sighting stick in a number of the projects to locate the exact position of objects you see in the sky. A two-axis device, the sighting stick allows you to measure altitude (the angle between an object and the horizon) and azimuth (the angle on the horizon between true north and a point directly below the object) with relative accuracy.

To build the sighting stick you will need a hand drill with a ¼-inch bit, a saw (a hand saw will do, but a hand jigsaw will make things easier), white glue, an X-Acto knife or single-edge razor blade, at least 12 inches of medium-weight thread, and two copies of the altitude and azimuth scales, which you can cut out from pages 116–119. Wooden parts needed are 10 inches of ¼-inch diameter doweling, a ½- to 1-inch thick flat board for the base, and a solid wooden thread spool or a 1½-inch piece of dowel about 2 inches long. For the sighting stick scale backs you will need two pieces of stiff cardboard. Save a large scrap of the cardboard to make a pointer for the azimuth scale and find two map pins to help hold a couple of things together.

Use one copy of the altitude and azimuth scales as patterns for cutting the cardboard to the right shape. Follow the outline of each carefully. Cut out and glue the second copy of each scale to the matching piece of cardboard. Measure the distance from the center of the azimuth (full circle) scale to the graduated circle and cut a triangle of cardboard whose base can completely cover the center hole while the "arrow"

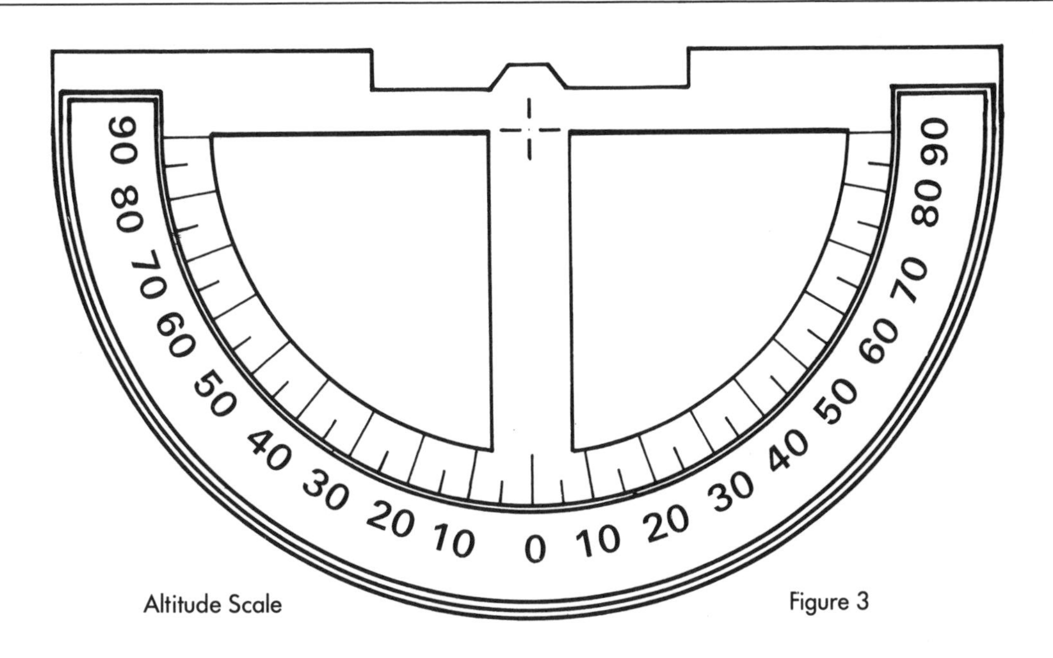

Altitude Scale

Figure 3

tip reaches the numbered circle. Next cut the ½- to 1-inch thick board into a 1-foot by 1-foot piece to make the base. In the center of this piece of wood, drill a ¼-inch hole. Cut a 1-inch piece of the ¼-inch diameter dowel. Spread glue onto one end of it and insert it into the hole in the base. Make a ¼-inch hole at the base of the "arrow" of cardboard cut previously and another ¼-inch hole in the center of the azimuth scale. If necessary, also bore out the hole at the center of the thread spool to ¼ inch or drill a ¼-inch hole down the center of the larger dowel if you do not have a spool. Glue the azimuth scale to the base board after placing the scale over the dowel at its center. Once these are dry, put the spool (with pointer attached) onto the dowel with the pointer directly against the azimuth scale. Do not glue the spool to the base dowel.

The altitude scale indicates in degrees how high an object is above the horizon. Cut-outs of this scale, which you will need to construct the sighting stick, are on pages 116–119.

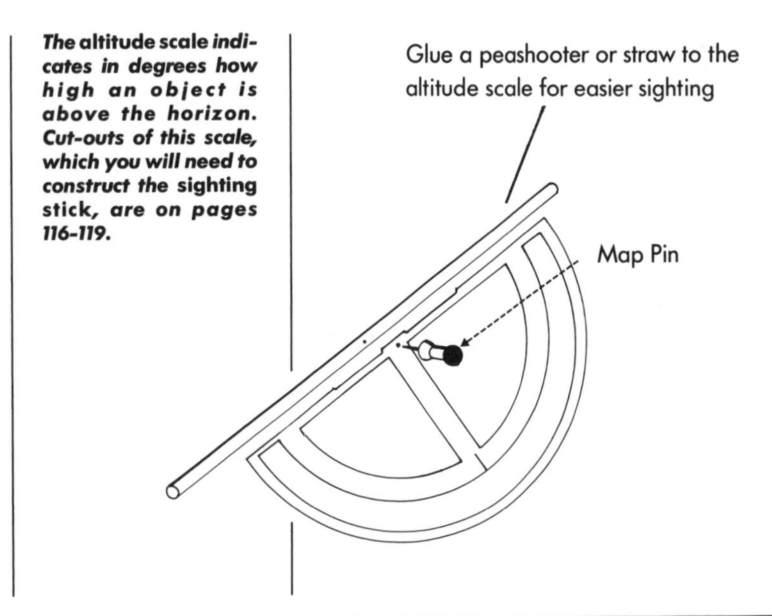

TOOLS OF THE TRADE

Place about 8 inches of the ¼-inch diameter doweling into the top of the spool. Mark the dowel where it meets the spool with a pencil. Take the spool off the base, apply glue to the longer dowel and replace it in the top end of the spool up to your mark. Let this dry before remounting onto the base. Once dry, use a map pin to mount the altitude scale to the top of the longer dowel. Tie one end of the thread to the map pin and the other end to a second pin or a fish weight. This weight will pull the thread down, allowing it to act as a pointer for the altitude scale. Place the spool back on the dowel

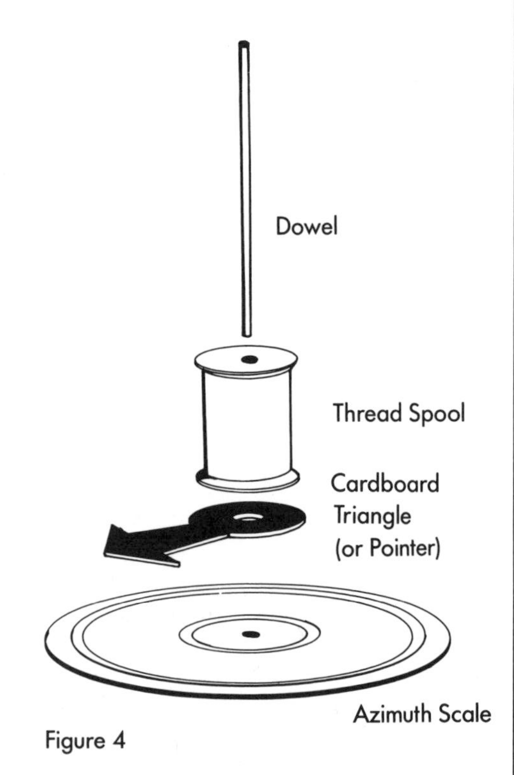

Figure 4

The azimuth scale indicates in degrees an object's distance from true north. Cut-outs of this scale, which you will need to construct the sighting stick, are on pages 116-119.

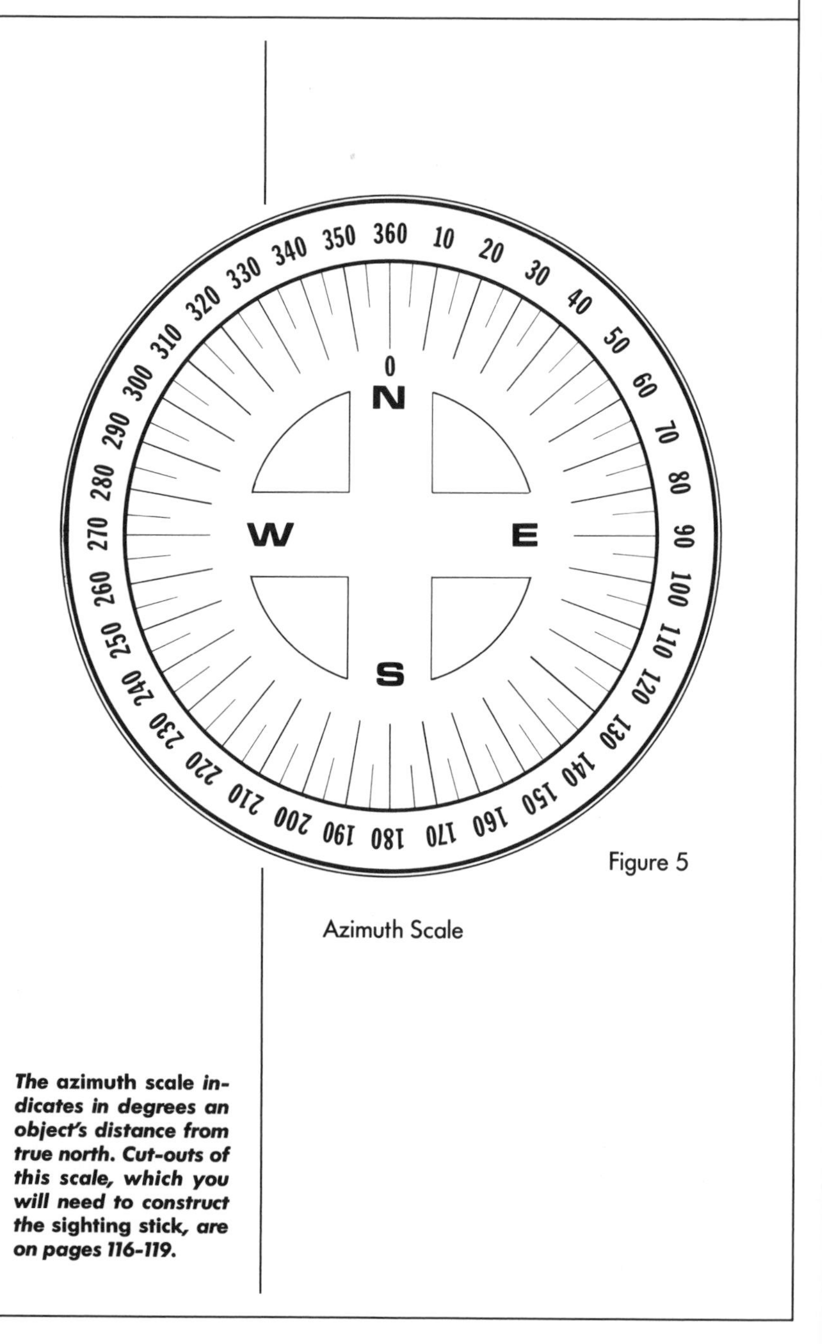

Figure 5

Azimuth Scale

that is mounted to the base and the sighting stick is ready for use.

With your sighting stick built, and after gathering any other necessary materials and any optical instruments needed, you are ready to do the projects in this book and to open up an exciting new world of learning. The only thing left to do is to master your physical orientation to the sky—the ability to align yourself, and your observing instruments, to the North Star so that your measurements and observations will make sense and be useful. The following exercise should help.

FINDING TRUE NORTH AND ALIGNING THE SIGHTING STICK

To find true north and align the sighting stick, you will need the star maps (pages 20–23), a flashlight with a red filter, a pencil, a C-clamp, and an optical instrument if you choose to use one.

Setting up

Mounting possibilities for the sighting stick are shown in Figures 6 and 7. You may want to try the alignment project that follows before deciding on a real mounting strategy since for tripod or stick mounting you will have to drill or nail the base of the sighting stick. For now, set the stick on a table or other flat surface and hold it in place with a C-clamp, if possible. Otherwise, be careful not to bump it for this exercise. Later you can try mounting it on a camera tripod, if you have one. An alternative is to attach the sighting stick to a post driven into the ground. Once you have decided what to place the stick on, be sure you set things up in a location that has a clear view of the sky, especially toward the north.

Figure 6

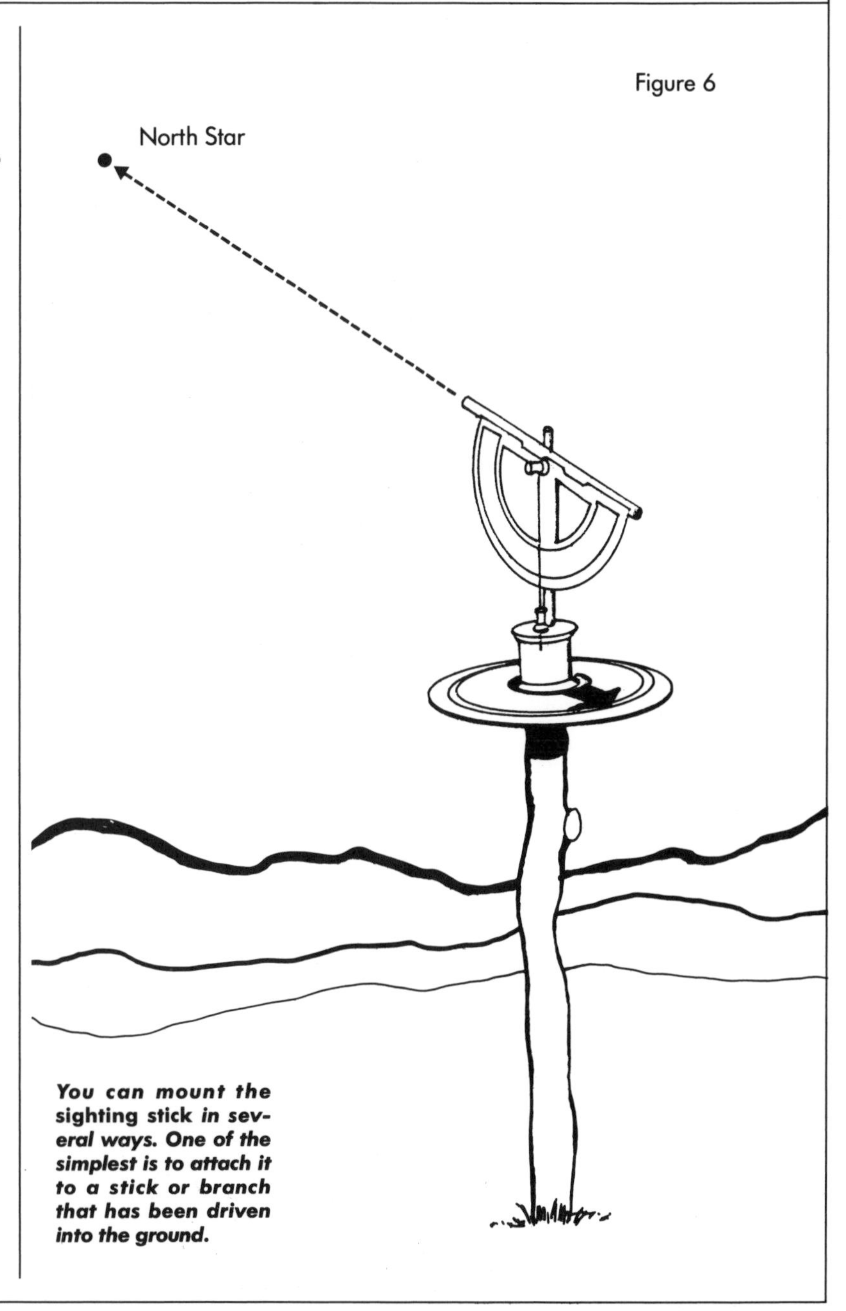

You can mount the sighting stick in several ways. One of the simplest is to attach it to a stick or branch that has been driven into the ground.

Finding true north and checking your latitude

There are several ways to orient yourself and your sighting stick to the north point on the horizon, but, unfortunately, the easiest—using a compass—is not the most accurate. Because the Earth's north magnetic pole does not coincide with the North Pole, from most places on our planet there is a difference, sometimes large, between true north and what a compass shows as north. The following method for finding true north will provide sufficient accuracy to complete the projects in this book—and it is easy. All you need is a clear, fairly dark sky, the star maps provided in this book, and your eyes. First, take a look at the star map (pages 20–23) for the current season and orient the map to the sky. The star maps are drawn for latitude 40 degrees north, but should be useful to skywatchers throughout North America, Europe, and Northern Asia. To align the maps to the sky, hold the map for the appropriate season in front of you, at one of the indicated times, with the direction you are facing at the bottom (i.e., if you are facing south, hold the map as shown on the page; if you are facing north, turn it upside down, etc.). To orient yourself to the sky, compare the stars you can see above the horizon with those indicated on the lower portion of the map. Don't forget that there may be planets in the sky that do not appear on the maps. If you can't find the Little Dipper, which includes the North Star, find the Big Dipper, which is visible in the sky almost all night during most of the year. An arrow on the star map for winter shows how to extend a line from the so-called Pointer Stars at the "front" of the cup of the Big Dipper (away from the handle) to the North Star. You can also refer to Figure 8 on page 24. You must find the North Star before moving on to the next step. You may want to practice this over a few nights before starting any other projects because, unless you can leave your sighting stick undisturbed from night to night, you will have to set it up each time.

Go outside with your star map and red-filtered flashlight and start getting adapted to the dark. Turn off the porch light or any other outside lights and try to get as far away as possible from houses and other sources of light. If necessary, find an area surrounded by enough trees to block the neighbors' lights. Unless you are in the middle of a city, you should be able to see the brightest stars. Of course, the best places are far from city lights, in the mountains, but for this project, you simply need a clear sky and no bright lights shining in your eyes. Be patient; it takes about fifteen minutes for most people to become completely dark-adapted. While waiting, compare your star map with any bright stars you see in the sky. You may not be able to see Polaris at first, because it is about forty-seventh on the brightness list. Look for the Big Dipper, Orion, or some other bright constellation as a starting point.

Once you begin to see some of the fainter stars, you can start looking for Polaris. Using the two Pointer Stars in the Big Dipper, extend a line from them about five times as long as the distance between them until you reach another star a little fainter than those in the Big Dipper. This star stands out in the area because it isn't near any bright stars. Using the star map, correctly oriented, try to trace out the Little Dipper—three stars in a curved handle, connecting to four in a rectangle shape. Once you have confirmed the location of the Little Dipper, you can be sure that you have found the North Star at the end of its handle. You may want to look for the W (or M) shape of the constellation Cassiopeia, opposite the Big Dipper. If the Big Dipper is not visible (as on late summer and early autumn evenings), try finding the Great Square of Pegasus and run a line along its western-most side to find the North Star (see Figure 9 on page 24).

Once you have found the North Star, you have located the north celestial pole as nearly as needed for our purposes here. Make sure your sighting stick is level with the true horizon. Use a carpenter's level if necessary.

When you have found the North Star, align the base of the sighting stick so that the 0-degree mark points toward you. Next, sight down the upper edge of the altitude scale and move the stick in both axes until you align the edge of the scale with the North Star. Look at the azimuth scale pointer. If it does not point to 0 when the stick is aligned to Polaris, carefully move the base until it does, making sure to keep the altitude scale edge in line with the North Star as you do so. You may need someone to help you do this, or you may have to do it in numerous small increments. Once the stick is aligned to the North Star and the azimuth pointer reads 0, clamp down the sighting stick. You have now aligned the azimuth scale for use in the projects.

Next, to check the accuracy of your altitude scale, look at the angle marked by the thread. Check this angle, if you can, against your latitude on a map (Rand McNally road atlases, among many others, have latitude and longitude markings in their margins). The angle marked by the thread and your latitude should be the same, or nearly so. Try to refine your measuring techniques to see if, with everything set up properly, you can almost precisely match the known latitude of your location.

By observing the North Star with your sighting stick, you have found your latitude and aligned the azimuth ring to north. It has been set up so that 0 degrees marks true north and everything else is measured from there. If you have to move the sighting stick between observations, you will have to realign it each time you observe.

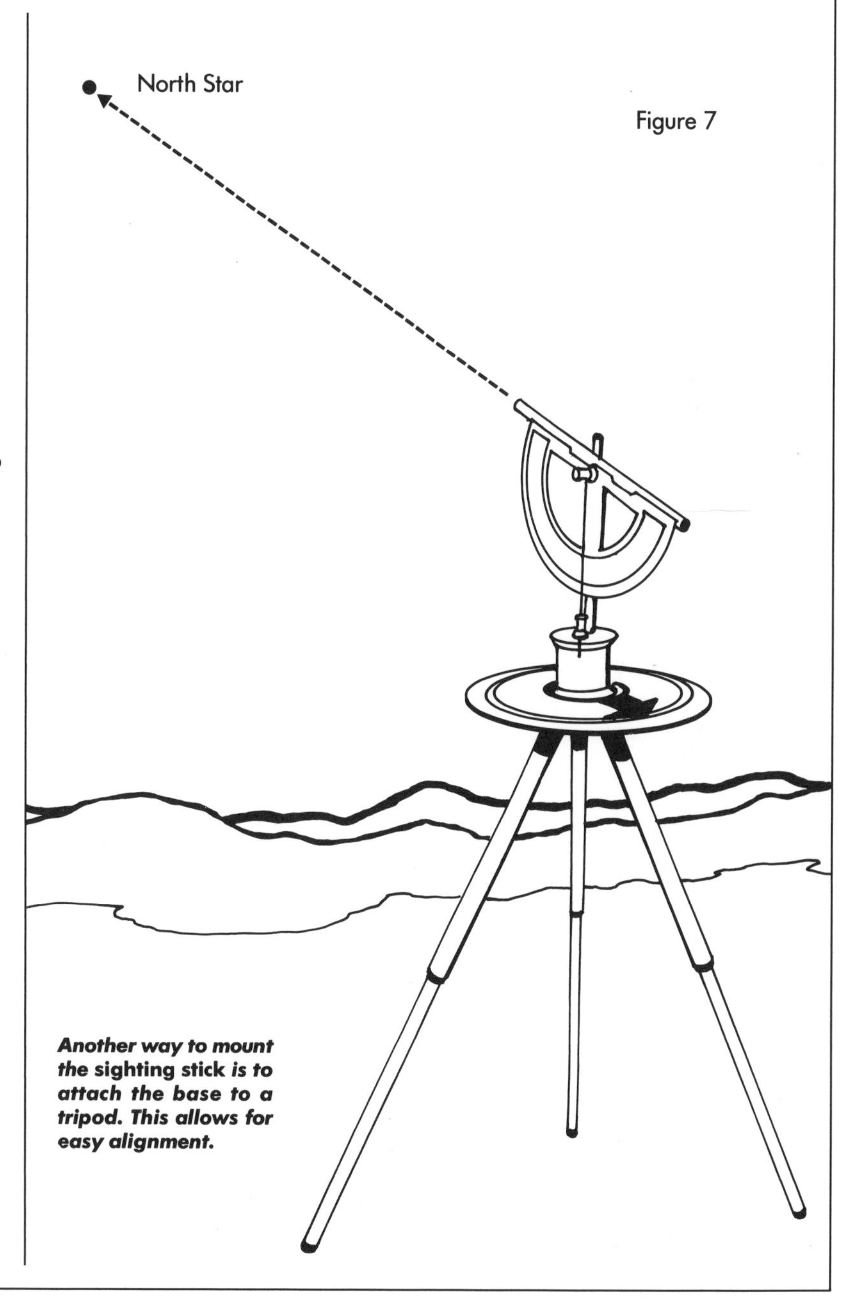

Figure 7

Another way to mount the sighting stick is to attach the base to a tripod. This allows for easy alignment.

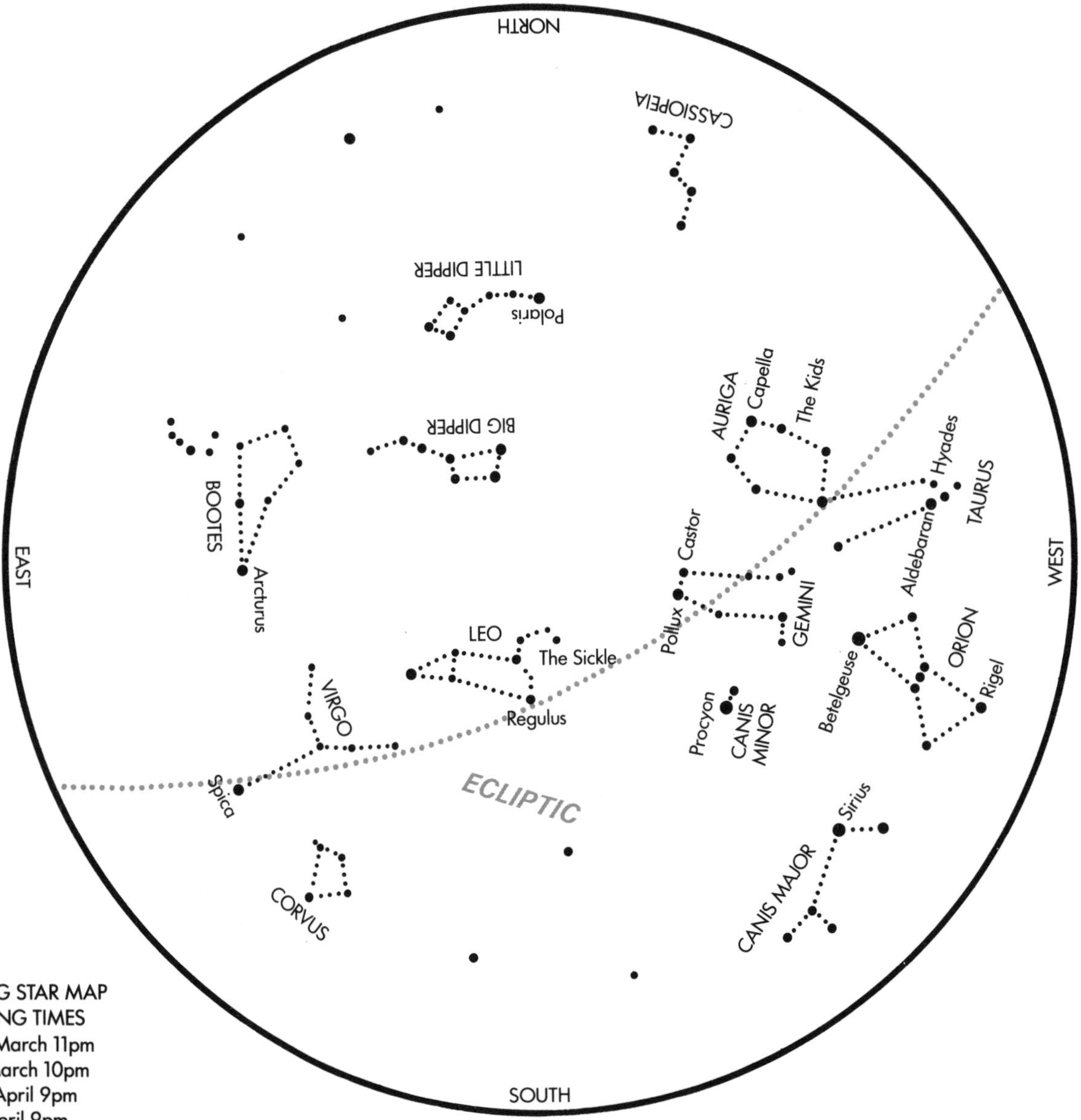

SPRING STAR MAP
VIEWING TIMES
Early March 11pm
Mid March 10pm
Early April 9pm
Mid April 9pm
(due to daylight saving time)
Early May 8pm

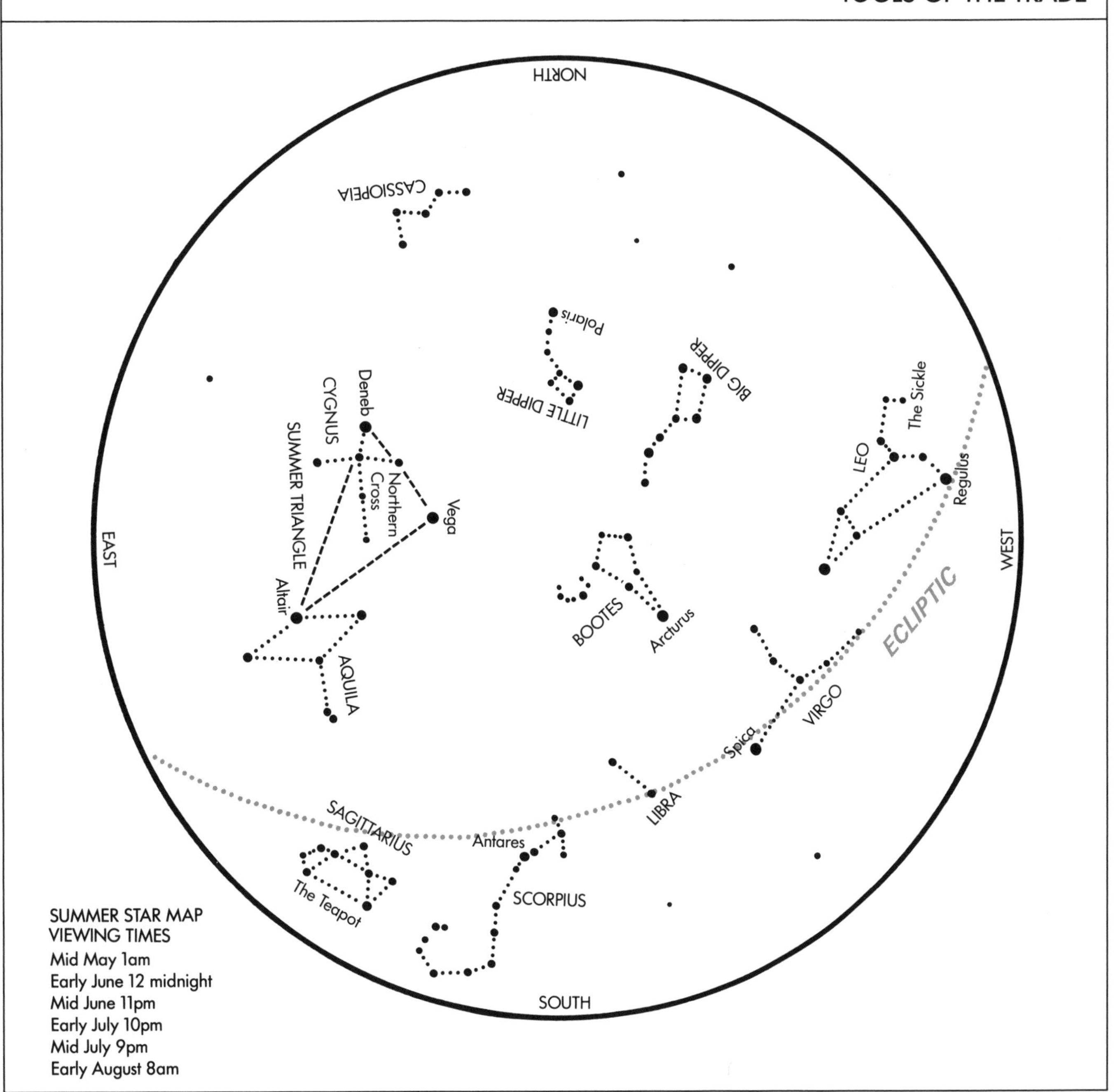
NORTH
SOUTH
EAST
WEST
CASSIOPEIA
Polaris
LITTLE DIPPER
BIG DIPPER
Deneb
CYGNUS
SUMMER TRIANGLE
Northern Cross
Vega
Altair
AQUILA
BOOTES
Arcturus
LEO
The Sickle
Regulus
ECLIPTIC
VIRGO
Spica
LIBRA
SAGITTARIUS
Antares
The Teapot
SCORPIUS
SUMMER STAR MAP
VIEWING TIMES
Mid May 1am
Early June 12 midnight
Mid June 11pm
Early July 10pm
Mid July 9pm
Early August 8am

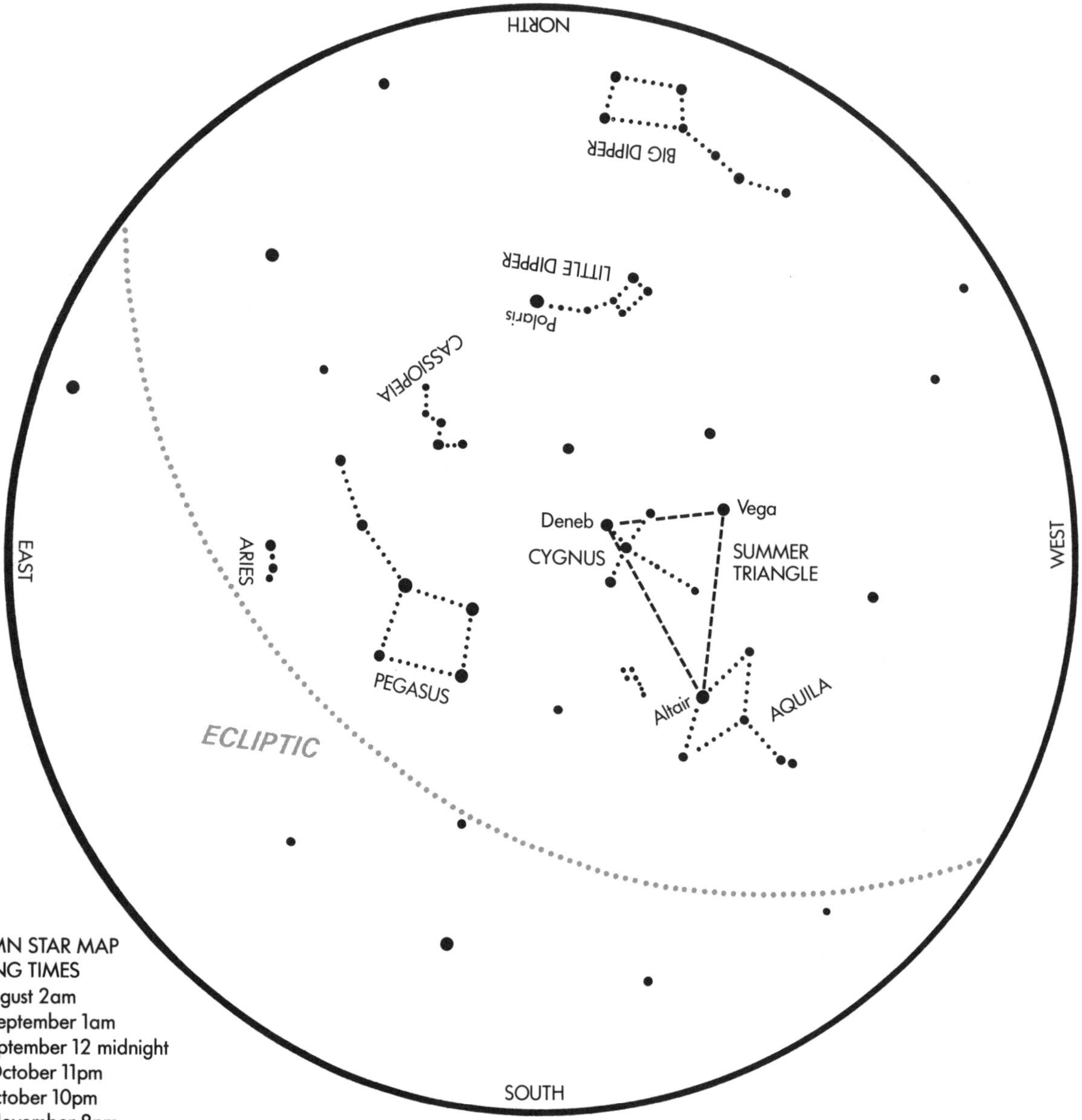

AUTUMN STAR MAP
VIEWING TIMES
Mid August 2am
Early September 1am
Mid September 12 midnight
Early October 11pm
Mid October 10pm
Early November 8pm
(due to return to standard time)
Mid November 7pm

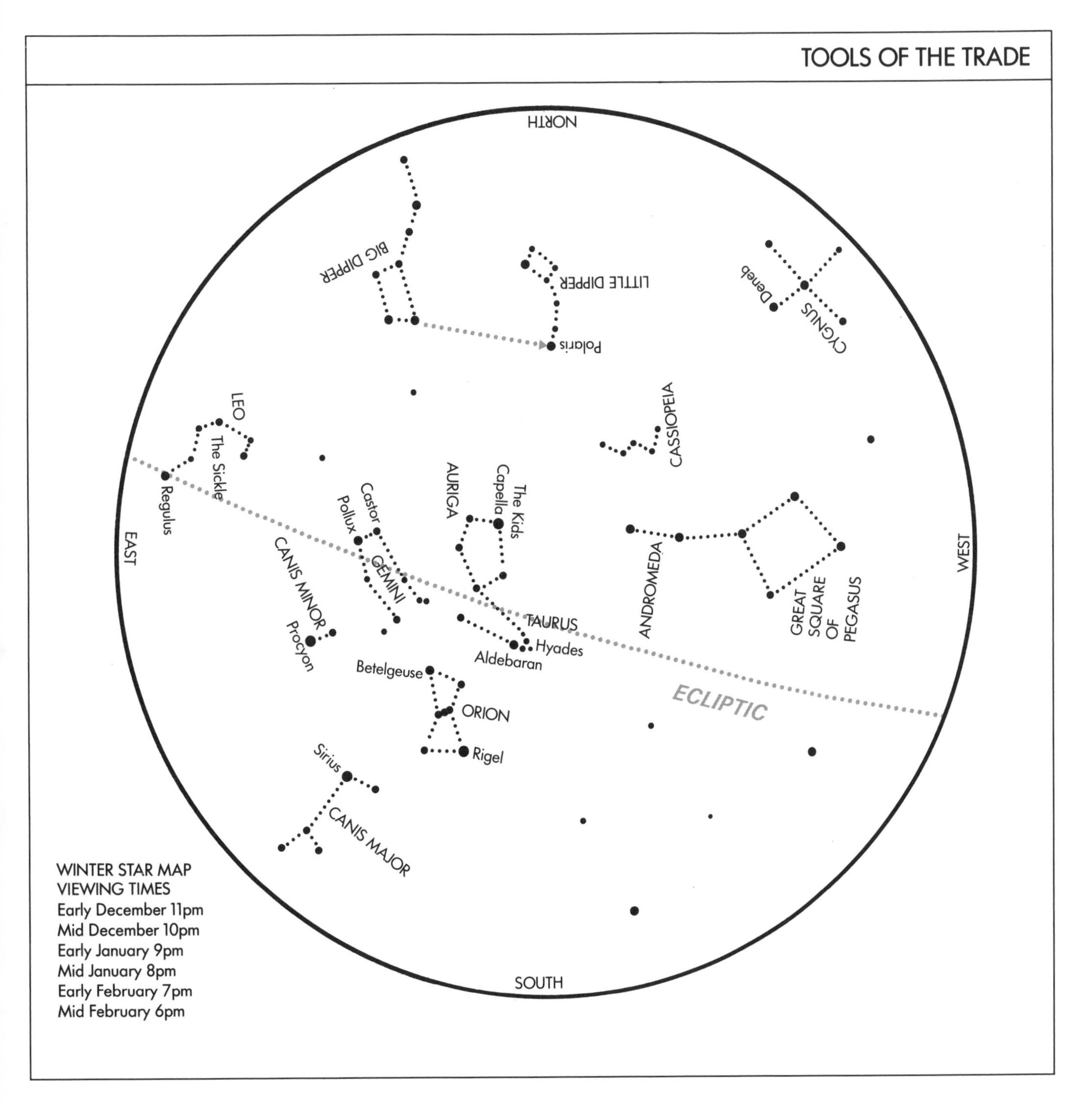
NORTH
BIG DIPPER
LITTLE DIPPER
Polaris
Deneb
CYGNUS
CASSIOPEIA
LEO
The Sickle
Regulus
EAST
WEST
SOUTH
AURIGA
Capella
The Kids
Castor
Pollux
GEMINI
CANIS MINOR
Procyon
ANDROMEDA
GREAT SQUARE OF PEGASUS
TAURUS
Hyades
Aldebaran
Betelgeuse
ORION
Rigel
ECLIPTIC
Sirius
CANIS MAJOR
WINTER STAR MAP
VIEWING TIMES
Early December 11pm
Mid December 10pm
Early January 9pm
Mid January 8pm
Early February 7pm
Mid February 6pm

Before using the sighting stick, you must align the azimuth scale to true north. You can use the Big Dipper to find Polaris, the North Star. Draw an imaginary line along the two stars farthest from the Dipper's handle and extend it out until it reaches the first bright star.

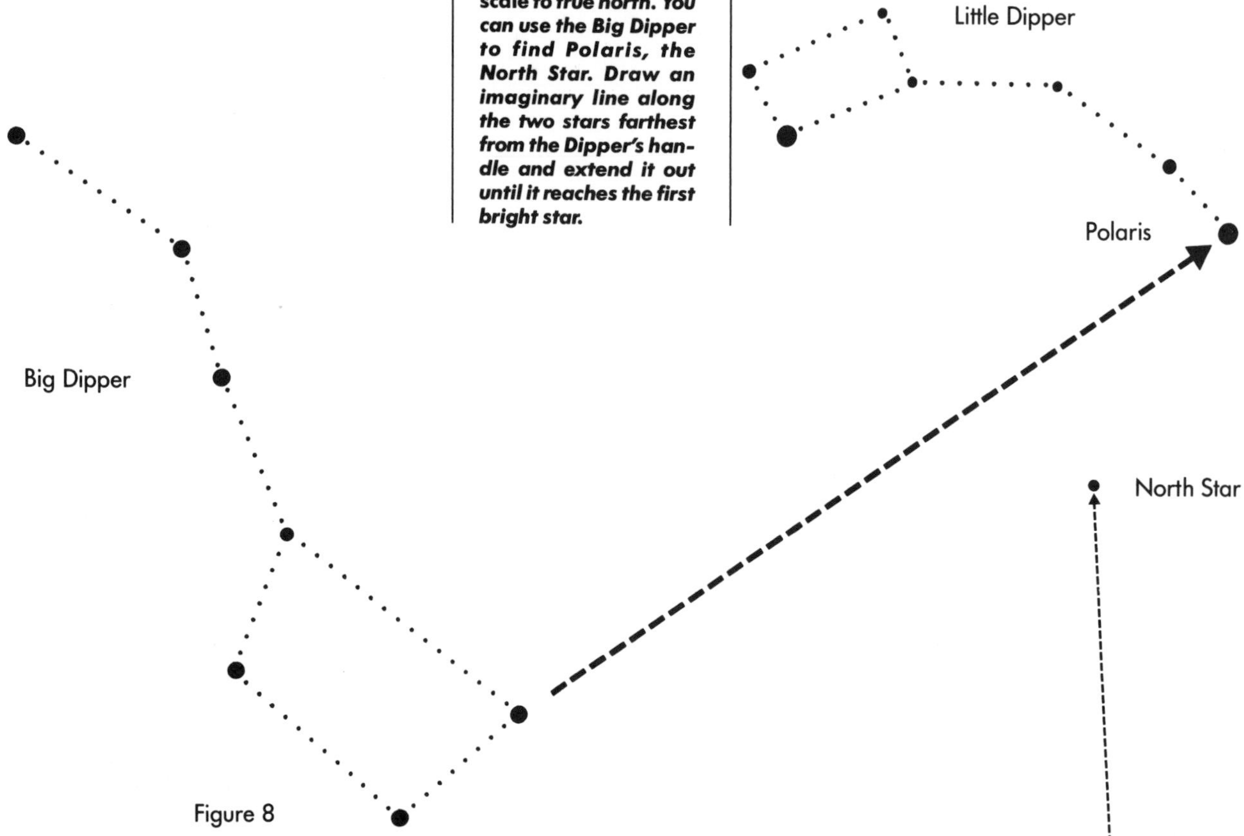

Little Dipper

Polaris

Big Dipper

Figure 8

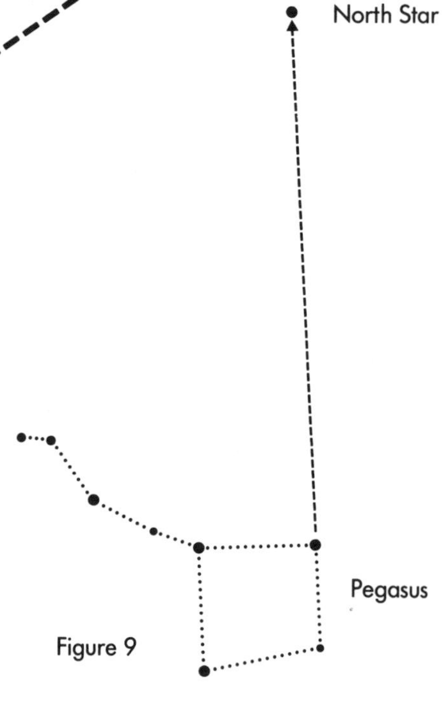

North Star

Pegasus

Figure 9

If the Big Dipper is not visible, as in the autumn, use the eastern side of the Great Square of Pegasus to find the North Star.

OPTICAL EQUIPMENT

The more advanced projects do require a telescope to carry out fully, but even some aspects of these will be interesting without one. If you are interested in purchasing optical equipment, here are some guidelines to follow to ensure a wise purchase.

Binoculars

One of the easiest and most versatile pieces of optical equipment is a pair of binoculars.

The first—and usually the least expensive—piece of equipment you'll want to buy is a pair of binoculars. If you are a bird-watcher or regularly attend sporting events, you probably already have these. Binoculars are marked with numbers that indicate *magnifying power × diameter of the lens.* Magnifying power tells you how many times larger the object will appear with the binoculars. The most common binoculars are 7 × 35 to 8 × 40, meaning 7- to 8-power magnification with 35- to 40-millimeter lenses. There are many binoculars on the market that are suitable for the kind of observation that will be done in conjunction with the projects in this book. For a good amount of light-gathering power and enough magnification to complete the Moon projects, an average pair of 7 × 35 binoculars or better is suggested. But there are binoculars that are a little smaller than 7 × 35's that can work well. If you have such a pair, try them before going out to buy a larger pair. The light-gathering power of the 35mm lenses, however, is highly preferred.

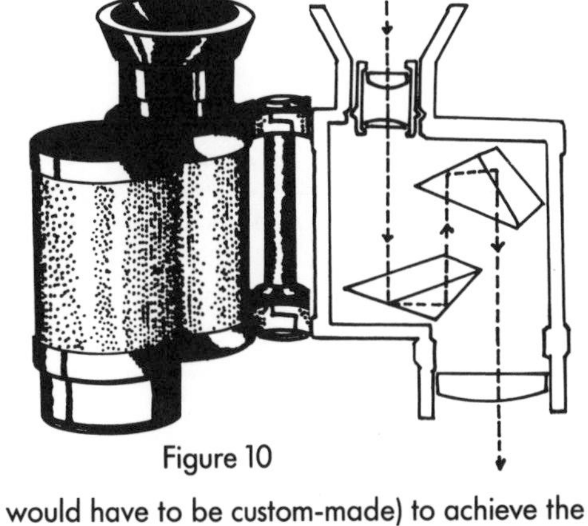
Figure 10

Spotting Scopes

A spotting scope is a small portable telescope that can be hand-held or mounted on a tripod.

Because they have a little more power and are easily mounted to a camera tripod, spotting scopes have several advantages over binoculars. First, spotting scopes offer higher magnification and normally have a zoom eyepiece that allows a continuous increase in power from 15X to 45X or even 60X. Second, most binoculars require a special adapter (which in most cases would have to be custom-made) to achieve the hands-free use and steadiness a tripod can provide. With such an adapter attached, hand-held use of the binoculars may become cumbersome or impossible. The main disadvantage of spotting scopes is their price. They often cost more than a good telescope with comparable light-gathering power and more magnification. Still, they are portable and easier to use than telescopes. Moving up to a telescope will require additional maintenance and expertise on your part.

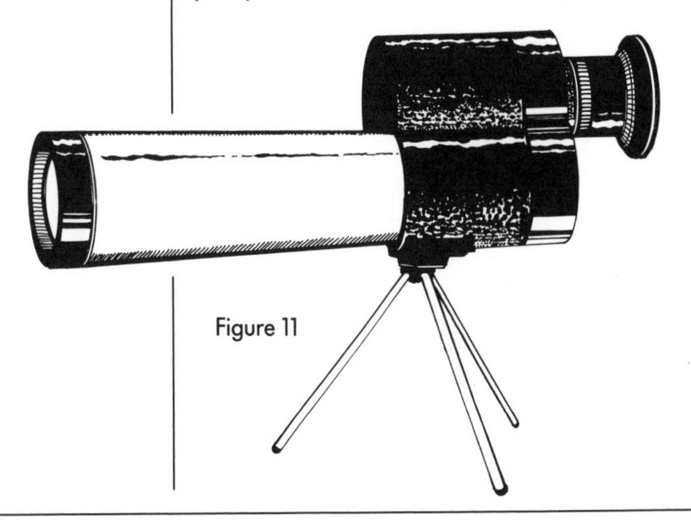
Figure 11

Telescopes

While telescopes provide the most varied and best combinations of magnifying power, lens diameter, and steadiness, choosing and using a telescope can be as frustrating (or rewarding) as buying a new car. Expertise among salespeople varies as widely as telescope styles. Don't buy a telescope simply because it has "high power." Keep in mind that the diameter of the lens (usually measured in millimeters) or mirror (measured in inches) tells you how much light you can gather, or how bright objects will appear. The diameter of the main lens or mirror, called the objective, and the quality of the optics determine how well you will be able to see an object—that is, how sharp it will appear. In almost all cases, you are better off seeing a clearer image at lower magnification. A good rule of thumb is that a magnification power of 50 is the maximum useful amount per inch of objective diameter. A higher ratio will cause the quality of the image to suffer. To quickly calculate the diameter of a lens in inches, divide the diameter in millimeters by 25 (60mm equals roughly 2.4 inches).

If a telescope has an equatorial mount (see Figure 13), its polar axis can be aligned to the North Star and the instrument can track the stars like an observatory telescope (Figure 12).

The best way to find out which telescope to buy is to call your nearest planetarium or amateur astronomy club. You may even be able to try out different instruments at a club gathering or a planetarium's public observing night. At the minimum, you should be able to speak to someone who can discuss the pros and cons of the telescopes you may be considering.

If you have a telescope with an equatorial mounting—that is, one that can be aligned to the north celestial pole (see Figure 13)—you may want to align it now if you are planning to use it. Most of the procedures that apply to aligning the sighting stick will be useful in

Figure 12

completing this task. You must find the North Star, and you can then align the polar axis of the telescope (shown in Figure 13) by sighting down the axis and roughly aligning it to Polaris. Some telescopes provide a sight for this purpose, making it quite easy. You may have to adjust the angle of the axis by using leveling bolts provided by the telescope manufacturer or by loosening a bolt or two on the mounting itself. Great care must be taken not to allow things to get out of balance either way. Read the manufacturer's instructions carefully for information on this procedure, or contact your local planetarium or astronomy club for help.

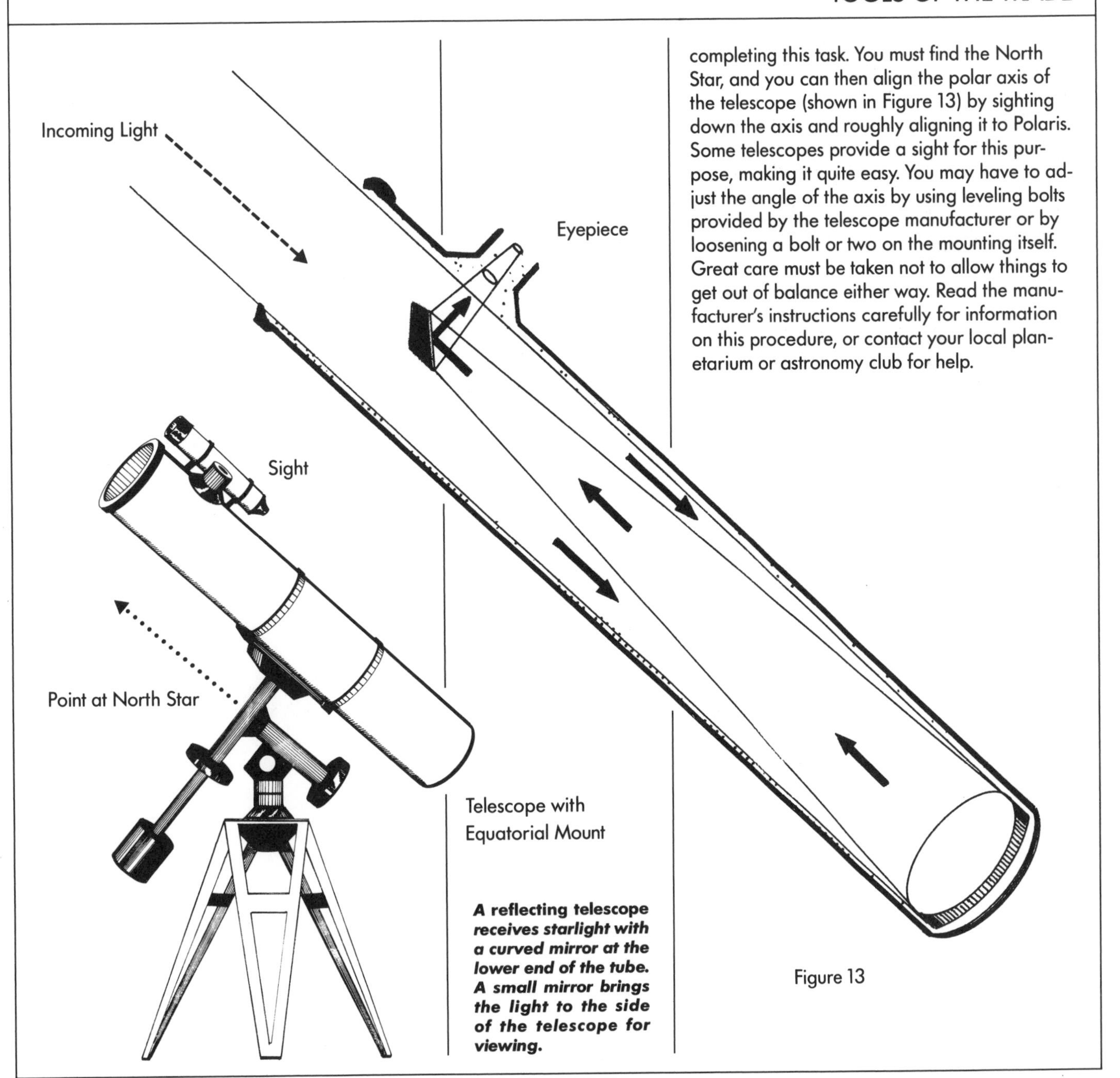

A reflecting telescope receives starlight with a curved mirror at the lower end of the tube. A small mirror brings the light to the side of the telescope for viewing.

Figure 13

Sun Viewer

In addition to the methods shown in illustrations in Project 5, there is another inexpensive but useful way to observe the Sun, if you have a telescope. This requires a little experimentation that must be carried out with extreme care. Always remember: NEVER LOOK AT THE SUN, especially through an optical aid. Also, do not use the good eyepieces from your telescope for this project. Nor should you use a spotting scope for this, because a spotting scope's eyepiece is usually built-in and expensive to replace.

After measuring the proper distance for focusing the Sun's image, cut out pieces of cardboard to construct a Sun viewer.

First, being very careful not to look into the eyepiece even from a distance, choose a sunny day and use the shadow of your telescope to line it up on the Sun. When the telescope tube's shadow looks as circular as it can be, you are lined up. Always use an inexpensive, lower-power eyepiece to project the Sun, since heat buildup during the observations can cause the glass to crack, and you do not want to ruin an expensive eyepiece! Once the telescope is lined up on the Sun, hold a white piece of cardboard about eighteen inches (45 cm) above the eyepiece and turn the focus knob until the disk of

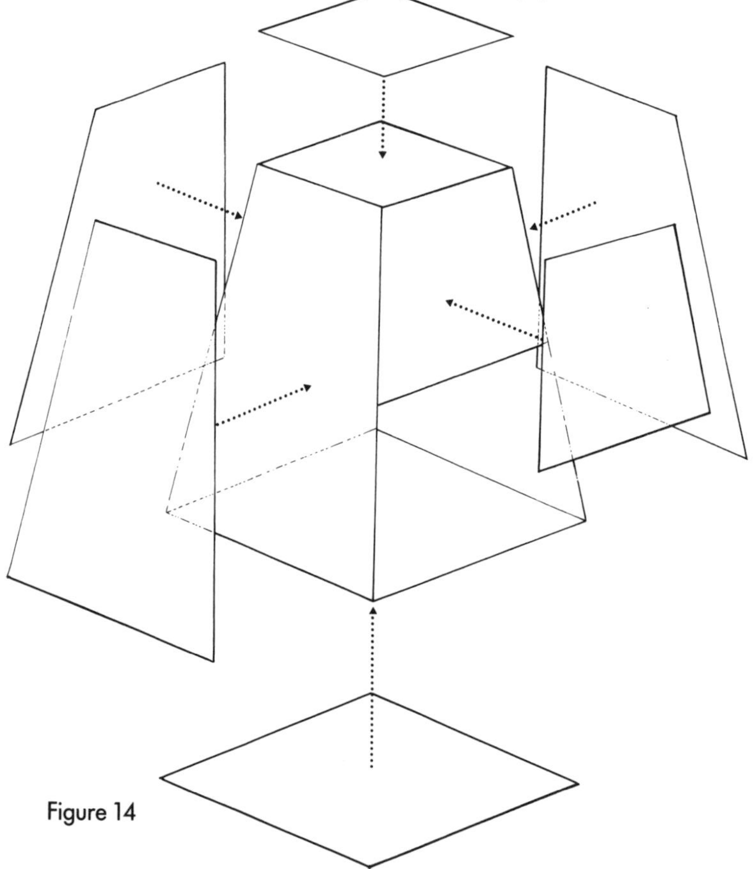

Figure 14

the Sun comes into focus. You will probably see sunspots on the cardboard that you can use to fine-focus on.

Do not let anyone touch the eyepiece or get in the light path between the eyepiece and the card. Measure the distance between the eyepiece and the card, and make sure that there is still a little travel in the focus mechanism to adjust for any errors. Also measure the diameter of the Sun image you are projecting. Now move the telescope so that it is no longer pointing at the Sun.

Now cut out three pieces of cardboard that are the length of the distance from the eyepiece to the card, and at one end a width two inches (5 cm) greater than the diameter of the Sun image you measured. At the other end, these pieces should taper down to about two inches (5 cm) wide.

Securely tape the three trapezoidal pieces together, then add the original square piece to the wider end. Now attach a tube (the cardboard tube from a toilet paper roll will do) to the smaller end and try the viewer out by attaching it to your eyepiece focusing mechanism. You may need to tape the viewer to the telescope to keep it secure. Rotate the viewer so that the open side is in a convenient position for viewing the projection of the Sun. Now realign your telescope to the Sun, and you should have a pretty good, shaded view of the Sun's projected image.

Attach a Sun viewer to the eyepiece mechanism of your telescope and arrange it so the image is easily and safely visible.

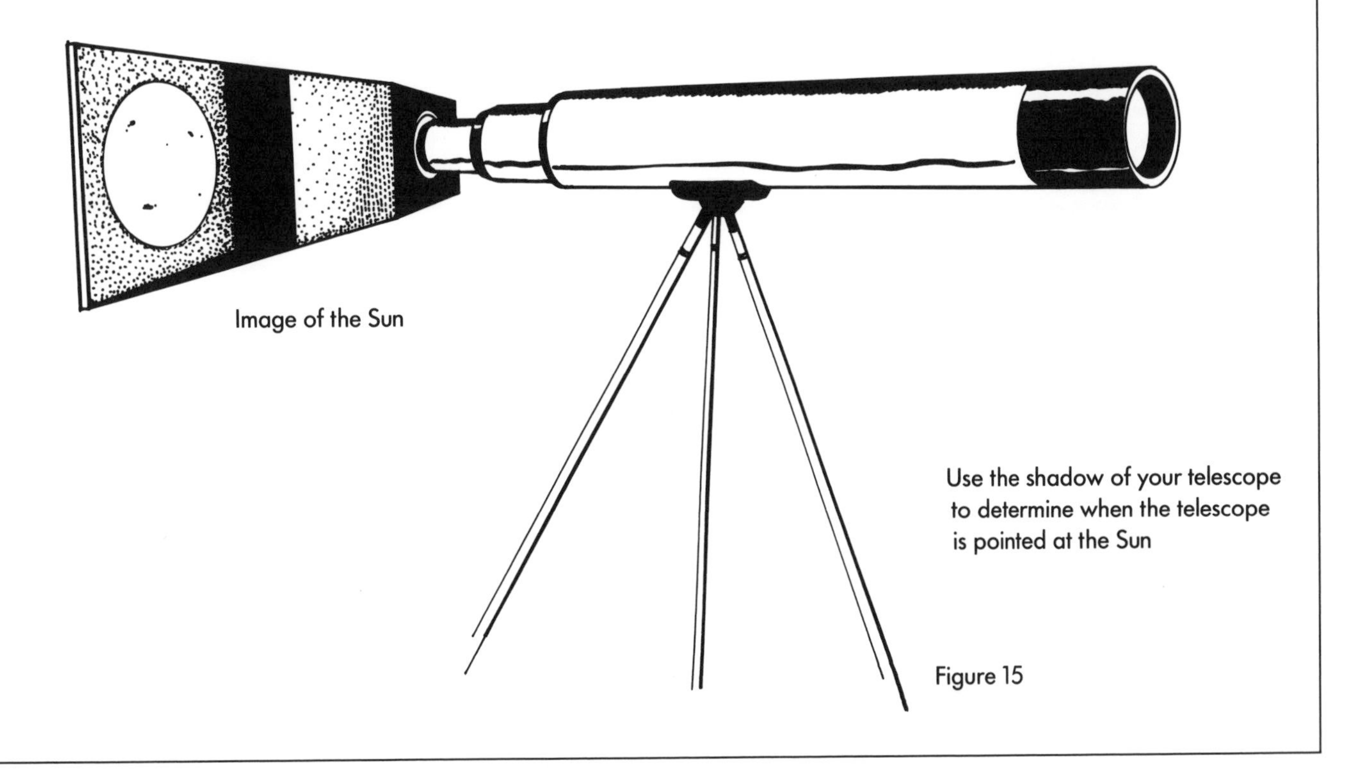

Figure 15

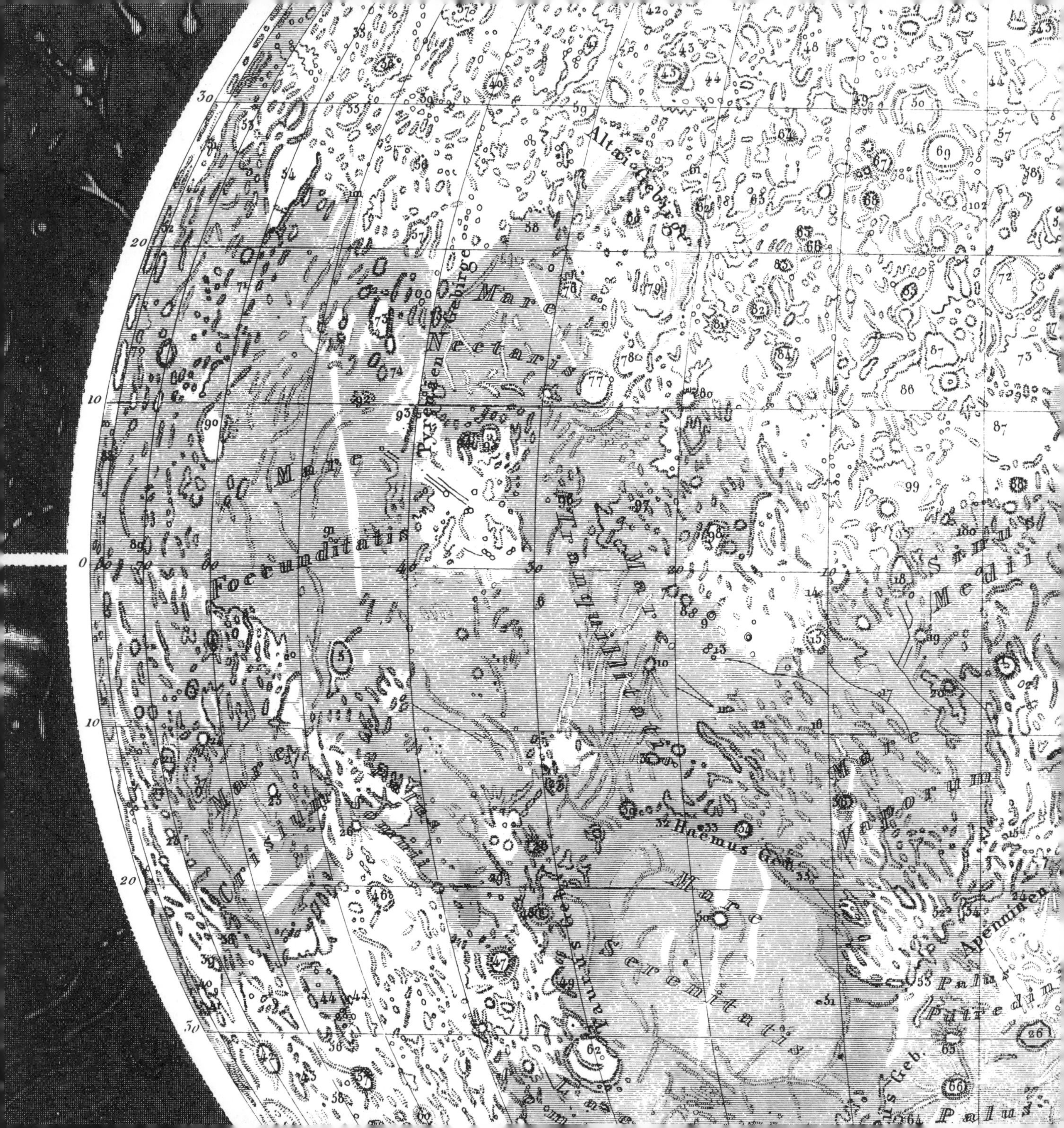

Altai Gebirge
Mare Nectaris
Pyrenäen Gebirge
Mare Foecunditatis
Mare Tranquillitatis
Sinus Medii
Mare Crisium
Haemus Geb.
Mare Vaporum
Mare Serenitatis
Apenninen
Palus Putredinis
Palus

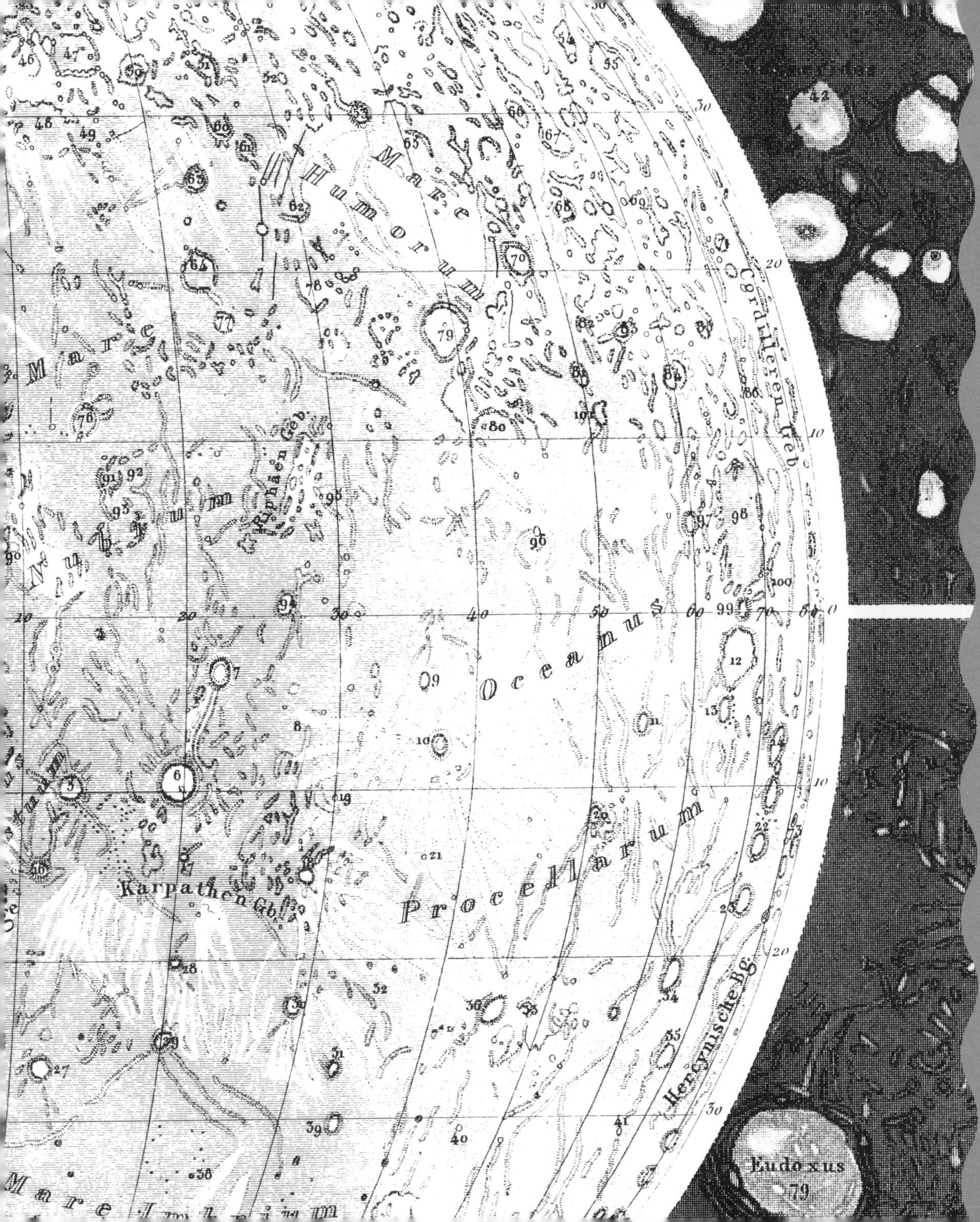

SECTION 2

MOON, SUN, STARS, AND ECLIPSES

1 PHASES OF THE MOON

LEVEL: Basic

EQUIPMENT: Observation form, pencil, sighting stick, watch or clock

PROJECT: Observe and record over time the altitude, azimuth and phases of the Moon

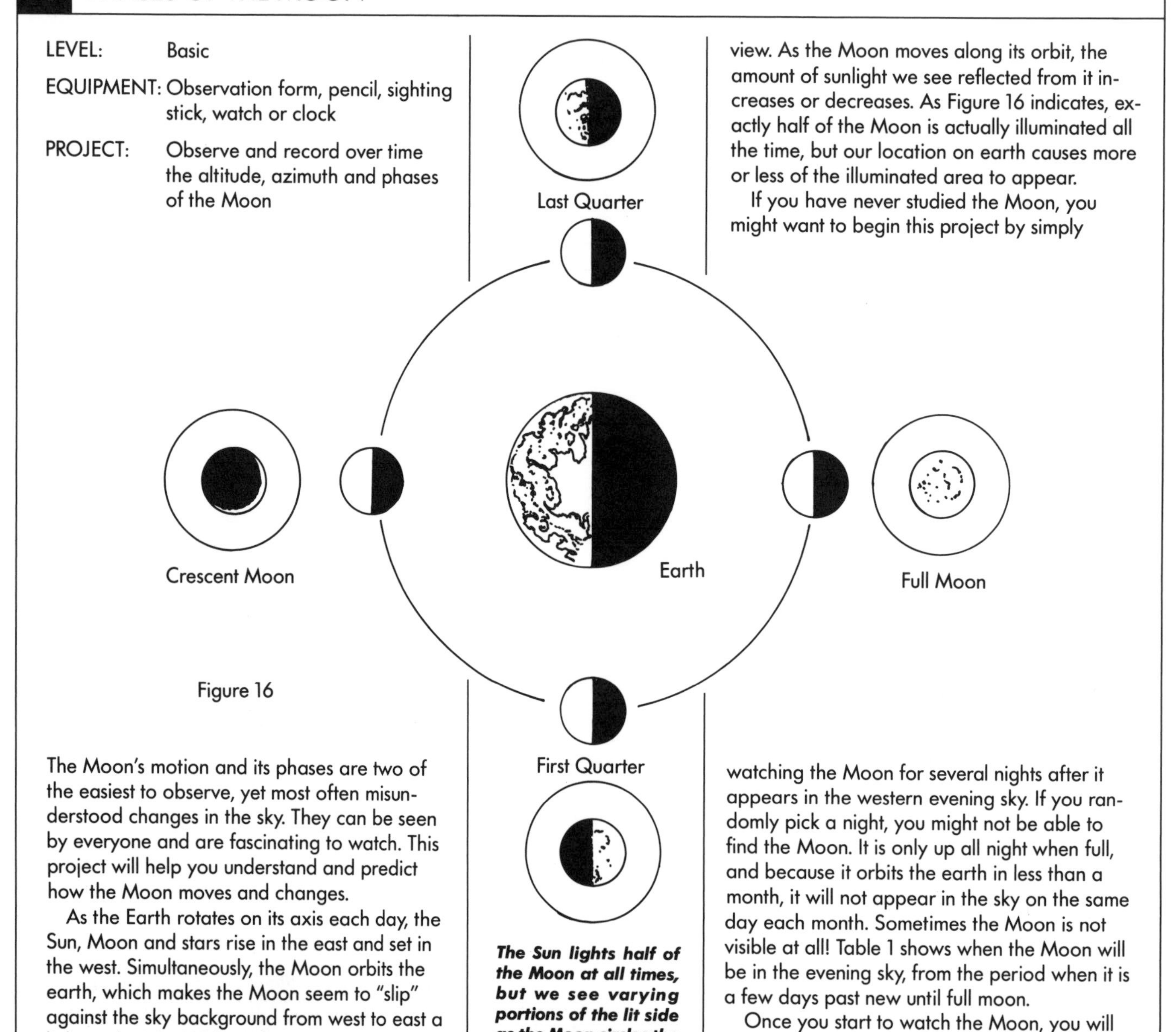

Figure 16

The Sun lights half of the Moon at all times, but we see varying portions of the lit side as the Moon circles the Earth. This variation causes the phases of the Moon.

The Moon's motion and its phases are two of the easiest to observe, yet most often misunderstood changes in the sky. They can be seen by everyone and are fascinating to watch. This project will help you understand and predict how the Moon moves and changes.

As the Earth rotates on its axis each day, the Sun, Moon and stars rise in the east and set in the west. Simultaneously, the Moon orbits the earth, which makes the Moon seem to "slip" against the sky background from west to east a little each night. With a telescope, you can see this motion by watching the Moon's position change relative to stars in the same field of view. As the Moon moves along its orbit, the amount of sunlight we see reflected from it increases or decreases. As Figure 16 indicates, exactly half of the Moon is actually illuminated all the time, but our location on earth causes more or less of the illuminated area to appear.

If you have never studied the Moon, you might want to begin this project by simply watching the Moon for several nights after it appears in the western evening sky. If you randomly pick a night, you might not be able to find the Moon. It is only up all night when full, and because it orbits the earth in less than a month, it will not appear in the sky on the same day each month. Sometimes the Moon is not visible at all! Table 1 shows when the Moon will be in the evening sky, from the period when it is a few days past new until full moon.

Once you start to watch the Moon, you will notice measurable changes almost immediately. Obvious changes should become apparent after two nights.

TABLE 1
PHASES OF THE MOON 1993–2002

n = new moon, q = first quarter, f = full moon, l = last quarter

	Jan	Feb	Mar	Apr	May	Jun	Jul	Aug	Sep	Oct	Nov	Dec
1993	8 f 14 l 22 n 30 q	6 f 13 l 21 n	1 q 8 f 14 l 23 n 30 q	6 f 13 l 21 n 29 q	5 f 13 l 21 n 28 q	4 f 12 l 19 n 26 q	3 f 11 l 19 n 25 q	2 f 10 l 17 n 24 q 31 f	9 l 15 n 22 q 30 f	8 l 15 n 22 q 30 f	7 l 13 n 20 q 29 f	6 l 13 n 20 q 28 f
1994	4 l 11 n 19 q 27 f	3 l 10 n 18 q 25 f	4 l 12 n 20 q 27 f	2 l 10 n 18 q 25 f	2 l 10 n 18 q 24 f 31 l	9 n 16 q 23 f 30 l	8 n 15 q 22 f 30 l	7 n 14 q 21 f 29 l	5 n 12 q 19 f 27 l	4 n 11 q 19 f 27 l	3 n 10 q 18 f 26 l	2 n 9 q 17 f 25 l
1995	1 n 8 q 16 f 23 l 30 n	7 q 15 f 22 l	1 n 9 q 16 f 23 l 30 n	8 q 15 f 21 l 29 n	7 q 14 f 21 l 29 n	6 q 12 f 19 l 27 n	5 q 12 f 19 l 27 n	3 q 10 f 17 l 25 n	2 q 8 f 16 l 24 n	1 q 8 f 16 l 24 n 30 q	7 f 15 l 22 n 29 q	6 f 15 l 21 n 28 q
1996	5 f 13 l 20 n 27 q	4 f 12 l 19 n 26 q	5 f 12 l 19 n 26 q	3 f 10 l 17 n 25 q	3 f 9 l 17 n 25 q	1 f 8 l 15 n 24 q 30 f	7 l 15 n 23 q 30 f	5 l 14 n 21 q 28 f	4 l 13 n 20 q 26 f	4 l 12 n 19 q 26 f	3 l 10 n 17 q 24 f	3 l 10 n 17 q 24 f
1997	1 l 8 n 15 q 23 f 31 l	7 n 14 q 22 f	2 l 8 n 15 q 23 f 31 l	7 n 14 q 22 f 29 l	6 n 14 q 22 f 29 l	5 n 12 q 19 f 27 l	4 n 12 q 19 f 26 l	3 n 11 q 18 f 24 l	1 n 9 q 16 f 23 l	1 n 9 f 15 f 22 l 31 n	7 q 14 f 21 l 29 n	7 q 13 f 21 l 29 n
1998	5 q 12 f 20 l 28 n	3 q 11 f 19 l 26 n	5 q 12 f 21 l 27 n	3 q 11 f 19 l 26 n	3 q 11 f 18 l 25 n	1 q 10 f 17 l 23 n	1 q 9 f 16 l 23 n 31 q	7 f 14 l 21 n 30 q	6 f 12 l 20 n 28 q	5 f 12 l 20 n 28 q	4 f 10 l 18 n 26 q	3 f 10 l 18 n 26 q
1999	1 f 9 l 17 n 24 q 31 f	8 l 16 n 22 q	2 f 10 l 17 n 24 q 31 f	8 l 15 n 22 q 30 f	8 l 15 n 22 q 30 f	7 l 13 n 20 q 28 f	6 l 12 n 20 q 28 f	4 l 11 n 18 q 26 f	2 l 9 n 17 q 25 f	2 l 9 n 17 q 24 f 31 l	7 n 16 q 23 f 29 l	7 n 15 q 22 f 29 l
2000	6 n 14 q 20 f 28 l	5 n 12 q 19 f 26 l	6 n 13 q 19 f 27 l	4 n 11 q 18 f 26 l	3 n 10 q 18f 26 l	2 n 8 q 16 f 24 l	1 n 8 q 16 f 24 l 30 n	6 q 15 f 22 l 29 n	5 q 13 f 20 l 27 n	5 q 13 f 20 l 27 n	4 q 11 f 18 l 25 n	3 q 11 f 17 l 25 n
2001	2 q 9 f 16 l 24 n	1 q 8 f 14 l 23 n	2 q 9 f 16 l 24 n	1 q 7 f 15 l 23 n 30 q	7 f 15 l 22 n 29 q	5 f 13 l 21 n 27 q	5 f 13 l 20 n 27 q	4 f 12 l 18 n 25 q	2 f 10 l 17 n 24 q	2 f 10 l 16 n 23 q	1 f 8 l 15 n 22 q 30 f	7 l 14 n 22 q 30 f
2002	5 l 13 n 21 f 28 q	4 l 12 n 20 q 27 f	5 l 13 n 21 q 29 f	4 l 12 n 20 q 26 f	4 l 12 n 19 q 26 f	2 l 10 n 17 q 24 f	2 l 10 n 17 q 24 f	1 l 8 n 15 q 22 f 30 l	6 n 13 q 21 f 29 l	6 n 13 q 21 f 29 l	4 n 11 q 19 f 27 l	4 n 11 q 19 f 26 l

1 PHASES OF THE MOON

It is interesting to note that, for some ethnic and religious groups, *new moon* means the first appearance of the Moon in the west, while astronomers define it as the instant the Moon passes closest to a point exactly between the Earth and the Sun. As shown in Figure 16, a day or two after new moon, as the Moon moves out of the solar glare, the thin waxing *crescent* becomes visible. Each following day, the Moon seems to grow larger as it heads toward full moon, which occurs almost two weeks after new moon.

A quick way to measure approximate angles is to hold your fist out at arm's length. The angle from one side of your hand to the other is approximately 10 degrees.

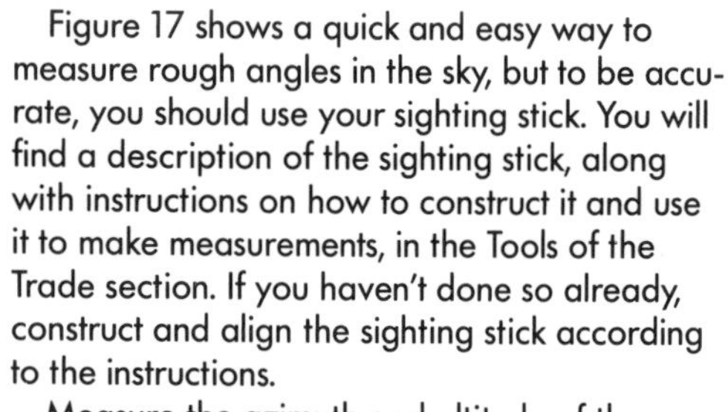

Figure 17 shows a quick and easy way to measure rough angles in the sky, but to be accurate, you should use your sighting stick. You will find a description of the sighting stick, along with instructions on how to construct it and use it to make measurements, in the Tools of the Trade section. If you haven't done so already, construct and align the sighting stick according to the instructions.

Measure the azimuth and altitude of the Moon when you first see it after new moon (See Figure 18). It is rare to see the Moon less than

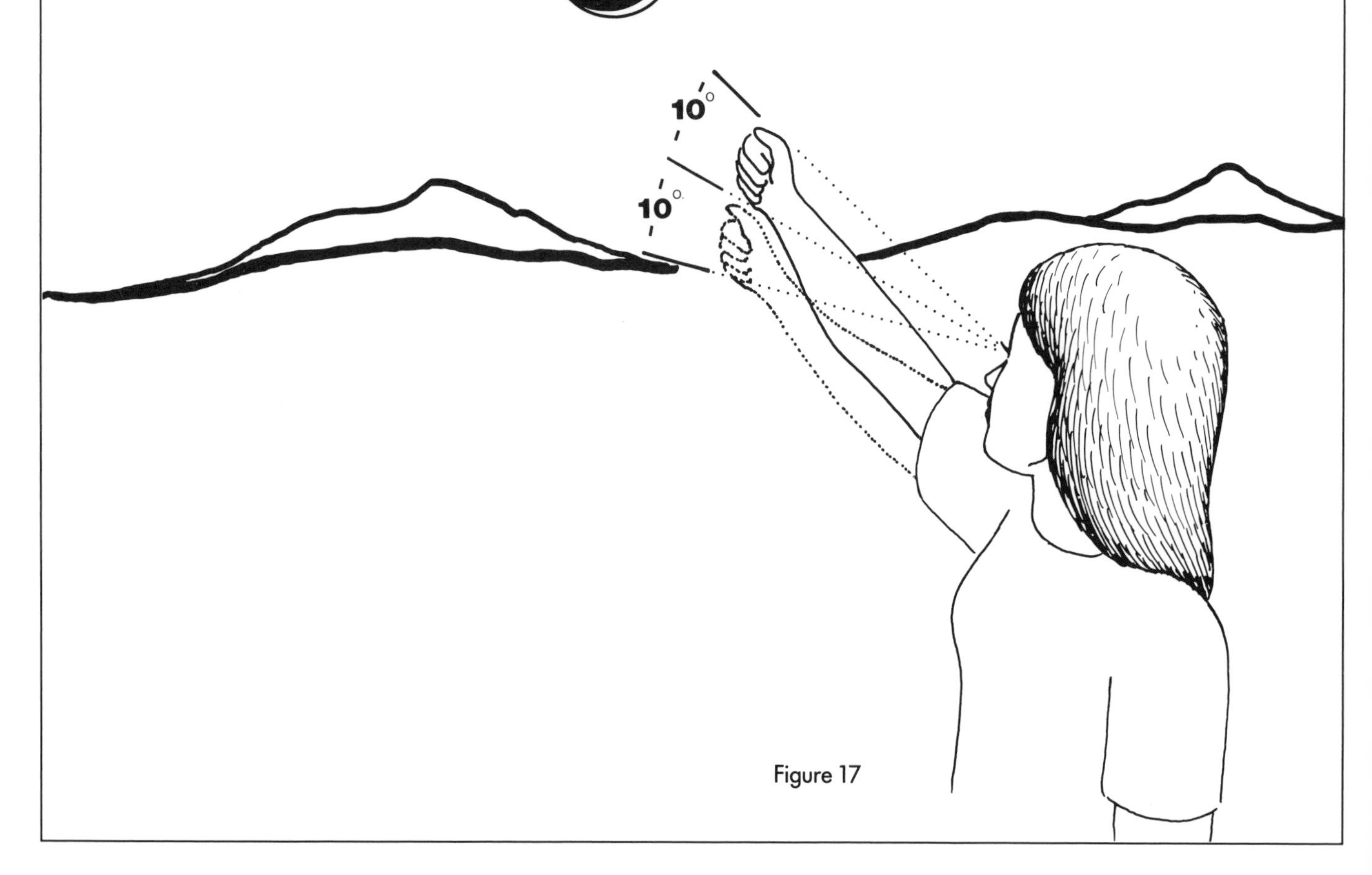

Figure 17

Figure 18

For more accurate measurement of angles, construct the sighting stick with the instructions provided in the Tools of the Trade section. This simple instrument will be useful for many of the projects.

twenty-four hours after new moon, but you might be able to spot a very thin crescent. Don't forget to record the time of your observation. With the Moon, things change very rapidly.

As you make this first observation, decide how often you want to observe and measure the Moon's phases and position. Observing every other night or every third night should allow you to see all the important changes. The key is observing at *regular intervals*. As time passes, the Moon seems to grow larger and larger until the angle formed by the Moon, Earth, and Sun is exactly 90 degrees. At this moment, the Moon has reached first quarter. Next comes the waxing gibbous phase, which leads to full moon, when the Moon is opposite the Sun in the sky. The rest of the cycle is the reverse of the first two phases. The gibbous phase wanes, and we reach last quarter, during which the Moon rises at about the middle of the night. Next comes the waning crescent moon, which lasts until the cycle starts again with another new moon.

As you make your observations, you may also want to use the sighting stick to measure the Moon's altitude and azimuth each night, or every other night, at *exactly* the same time. Record your observations on the Observation Form (page 9) for later reference. If you observe several whole lunar cycles, or the same part of several cycles, you will notice that the Moon's altitude and azimuth change somewhat even if you measure them during exactly the same portion of two different cycles. You may also want to record your observations on a copy of the star map (pages 20–23) for the season during which you are observing. This will help give you an idea of how the Moon follows a different path from the Sun and the planets in the sky. Finally, think about what happens if the Moon crosses the ecliptic (defined in the introduction) at the point where the Sun happens to be, or directly opposite that point. We will cover this in more detail in Project 5.

2 OCCULTATIONS OF PLANETS AND STARS

LEVEL: Basic

EQUIPMENT: Observation form, pencil, watch or clock, binoculars (telescope or spotting scope optional)

PROJECT: Time and record when planets or stars disappear and reappear as the Moon passes in front of them

Besides going through phases, the Moon appears to move rapidly against the background of stars in the sky when compared to all other astronomical objects except meteors. It orbits the Earth at a mere 250,000 miles (about 400,000 kilometers—a short distance in astronomical terms) and travels nearly 1.6 million miles (2.7 million kilometers) in less than a month. As it moves, it passes in front of objects in its apparent path from our point of view on Earth. Although the word *occultation* conjures up images of witches and spells, in astronomy it refers simply to the phenomenon of one object passing in front of and covering another. Figure 19 shows a lunar occultation of a star. Observing occultations is one way astronomers have determined the Moon's size and its distance from Earth.

Since the Moon appears relatively large in the sky, it covers a band as it makes its "backward" trek against the stars. If you carried out Project 1 fully or have studied the Moon often, you have probably noticed that it seems to appear higher in the sky at some times than at others. Because the Moon's orbit is tilted when compared to the plane of the Earth's orbit, the Moon can pass in front of a fairly wide range of stars, as well as all of the planets, over the course of time, as you can see in Figure 20.

Except at full moon, any time you view the Moon with binoculars, you will see faint stars near its dark edge. Some of these will have just come out from or will be about to disappear behind the Moon. In this project you will observe a star or planet, record the exact times it disappears behind the moon and reappears, and carefully note the exact location where you make your observations. To amass truly valuable occultation information, your time standard must be precise. If you have access to the National Bureau of Standards time signal channel, WWV, tune it in for this project. Many weather radios can receive this channel.

Start with occultations of bright objects. One of the most beautiful sights in the sky is a crescent moon near a bright star or planet. At such times, you may be lucky enough to observe an occultation of the object, depending upon your location. (The Moon is close enough to Earth, and the stars so far away, that where you are on Earth influences the occultations you will see.) Two good sources for finding out when many occultations (and especially major ones) are going to occur are the most recent January issue of *Sky and Telescope* magazine and *The Observer's Handbook* published by the Royal Astronomical Society of Canada. You will need one of these, or a similar source, to find out when occultations of bright objects will be visible from your area.

You should be able to observe major occultations with the unaided eye. If the object to be occulted is a star, you will see it "wink out" and should record the moment this happens on your Observation Form. When they are occulted, stars disappear instantly because they appear as points of light, but planets have discernable disks, so when they're occulted they will seem to fade out quickly. Actually, without optical aid, a planet's disappearance may also appear to be instantaneous, but with a telescope, the disk of the planet will be visible. Try to record the time when the "leading edge" of the planet begins to be covered (first contact), then when the entire planet is covered (second contact).

How long an object remains occulted by the Moon depends on what part of the lunar disk it

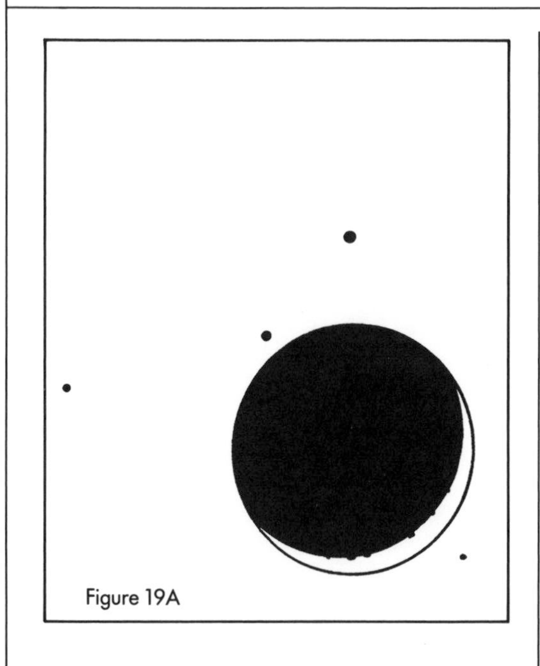
Figure 19A

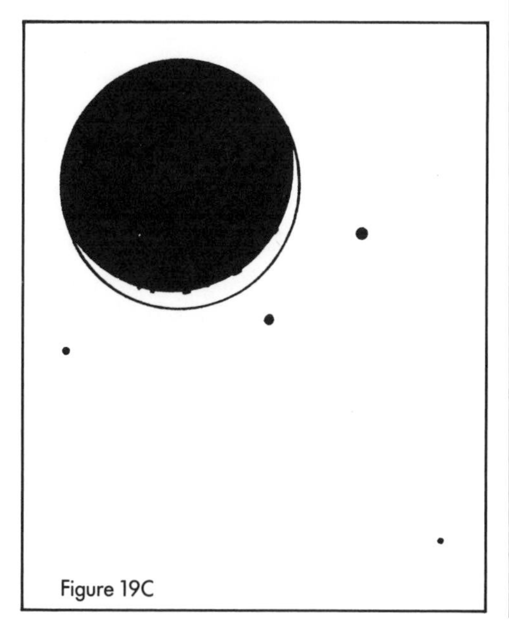
Figure 19C

As the Moon moves in its orbit, it passes in front of many stars and even planets, as seen from Earth. This phenomenon is called an occultation.

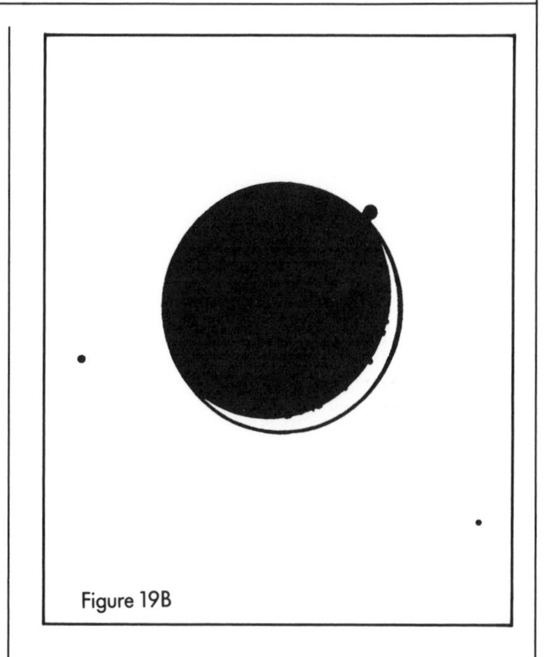
Figure 19B

Figure 19

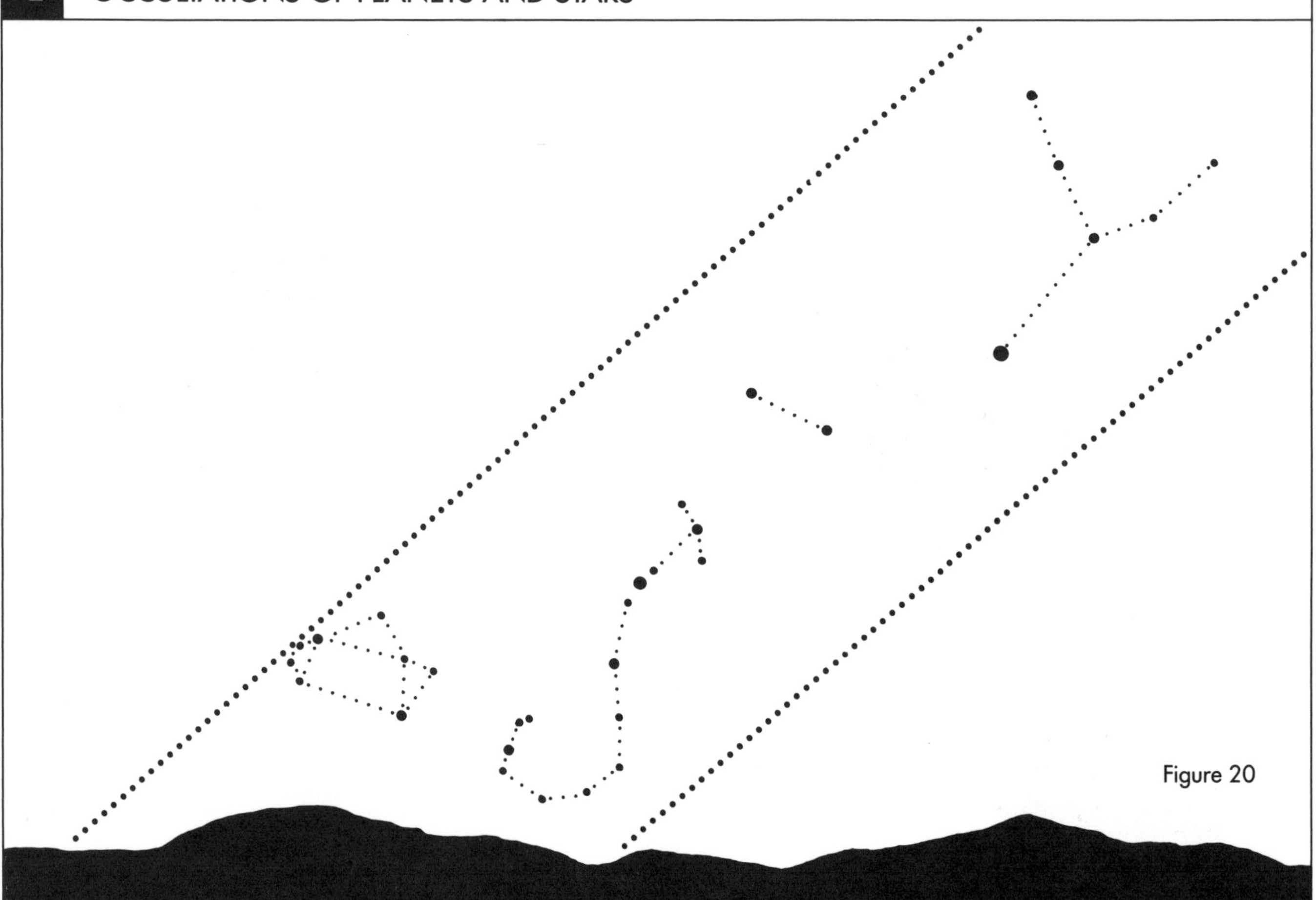

Figure 20

goes behind. Being ready for its reappearance requires some patience and a little research. Times from the previously mentioned references will help you approximate when the object should reappear. With planets, third contact—the time when the planetary disk begins to reappear—is often missed, but you may be able to catch it by carefully watching the edge of the Moon at the time and place you think the planet should reappear. Fourth contact is the time when the object has completely reappeared from behind the Moon.

Since the Moon's orbit is tilted relative to Earth's, our nearest neighbor in space can appear anywhere along a large band of the sky.

Whether you observe the occultation of a planet or a bright or dim star, record the appropriate exact times carefully. But make sure the object appearing from behind the Moon is the same one you timed disappearing. You can observe occultations of many stars, but only those you follow throughout the entire process will yield valuable information. Once you have gained experience in timing occultations, you can write to the Association of Lunar and Planetary Observers and offer your observations to contribute to that organization's research.

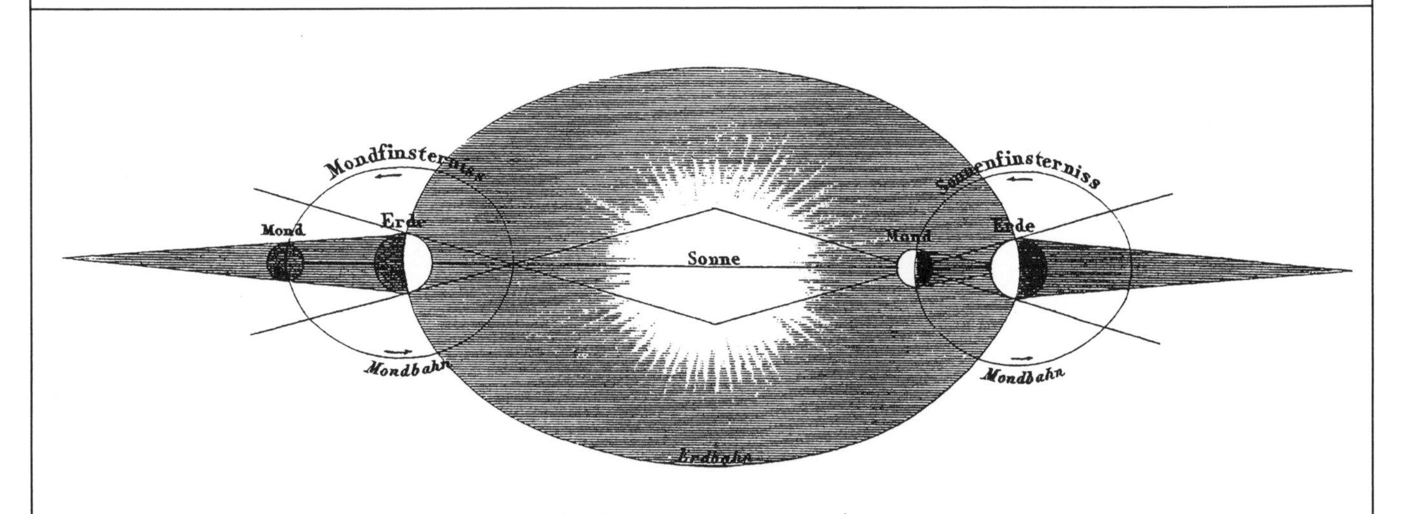

An occultation occurs when one celestial object passes in front of and covers another. An eclipse occurs when the Sun, Earth, and Moon become aligned in a way that blocks the Sun or the Moon from the view of observers on Earth.

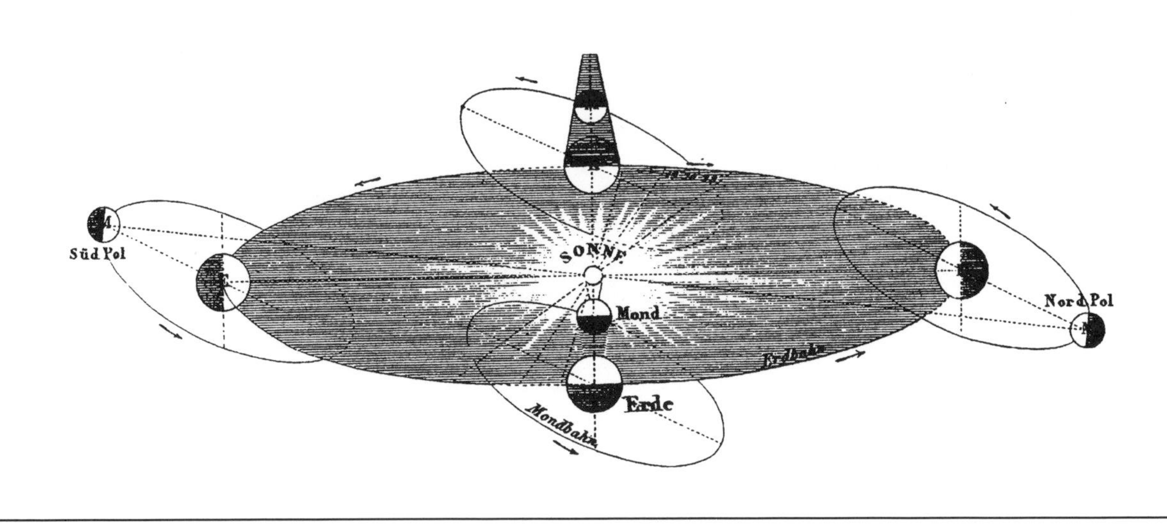

3 FEATURES OF THE MOON

LEVEL: Intermediate

EQUIPMENT: Telescope or spotting scope on mount or tripod, pencil, paper, Observation form

PROJECT: Observe and draw features on the Moon that are visible at medium to high power

Until the invention of the telescope, we thought the Moon had a much smoother surface than it actually does. Human beings had always noticed, though, that certain areas of the Moon's surface appear darker than others. Early observers thought that these dark areas were bodies of water, and so they called them *maria*, which means "seas" in Latin. Even today we call these dark areas "seas," though we now know that there is no water on the Moon. These "seas" are areas on the lunar surface where darker material from the interior of the Moon "boiled" up in a molten state (during some violent phase of the Moon's past) and solidified into huge beds of lava. The Moon's surface is also pitted with countless craters, visible with only 30- or 40-power magnification. These are evidence that the Moon has no atmosphere to protect it from objects bombarding it from space.

To begin this project, choose a good time to see lunar features. Different phases reveal different features better, because the *terminator*, the line where the lunar day and night meet, causes long shadows on the surface of the Moon. For an overall look, without concern for details, observing the full moon can give you a feeling for the general distribution of major features (see Figure 21). The darker, flatter areas seem to be clumped together, mainly in one hemisphere, while the higher, more mountainous, and cratered areas are concentrated in the other. Although you can see the entire Moon at full moon, there are no shadows to help highlight mountains, craters and valleys. To study the Moon in detail, choose another time. A good map of the Moon will enhance your enjoyment of this project, because you can use it to check which features you are seeing as the lunar phases progress. From about three days after new moon until three days before full, more and more features become visible each day (see Figure 23). Table 2 provides a brief list of major features and how and when they can best be observed in the first half of the lunar cycle. For a day-by-day description, maps and full information, consult the detailed description in *The Moon Observer's Handbook*, by Fred W. Price.

At full moon, you can see the general distribution of lunar features, but to observe the Moon in detail, look near the terminator, the line where lunar day and night meet, during gibbous or crescent phases. The Moon displays a variety of features from high mountain areas and craters to flat lava beds and meandering valleys.

TABLE 2
Features of the Moon and When They Are Visible

Days past new moon	Prominent craters and other features
4	Langrenus, Geminus, Hercules, Atlas
6	Aristotle, Maurolycus, Azophi, Abenezra
8	Clavius, Tycho, Copernicus, Plato
10	Scheiner, Jura Mountains, Longomontanus
12	Grimaldi, Darwin

To begin your detailed observation, scan the Moon's surface with your telescope (or spotting scope), starting from the brightly lit edge and moving across toward the terminator. Craters, mountains and valleys will stand out starkly near the terminator but elsewhere they'll blend in more with their surroundings. Try to find different features, or even different kinds of craters. Some have bowl-shaped bottoms, while others have central peaks. Some craters have smaller craters within them or along their slopes or rims. Some have even been partially "washed away" by molten material during a more active

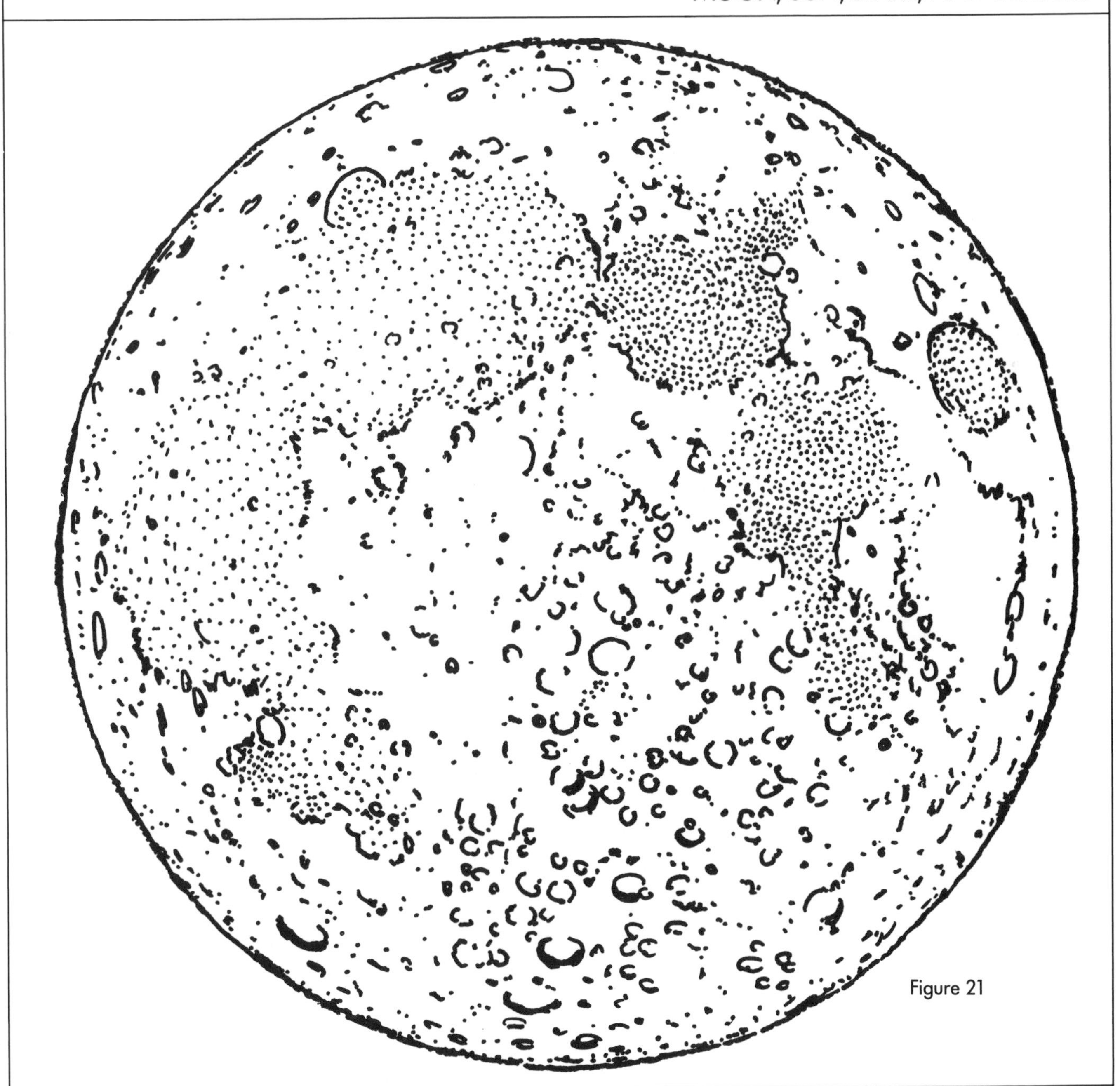

Figure 21

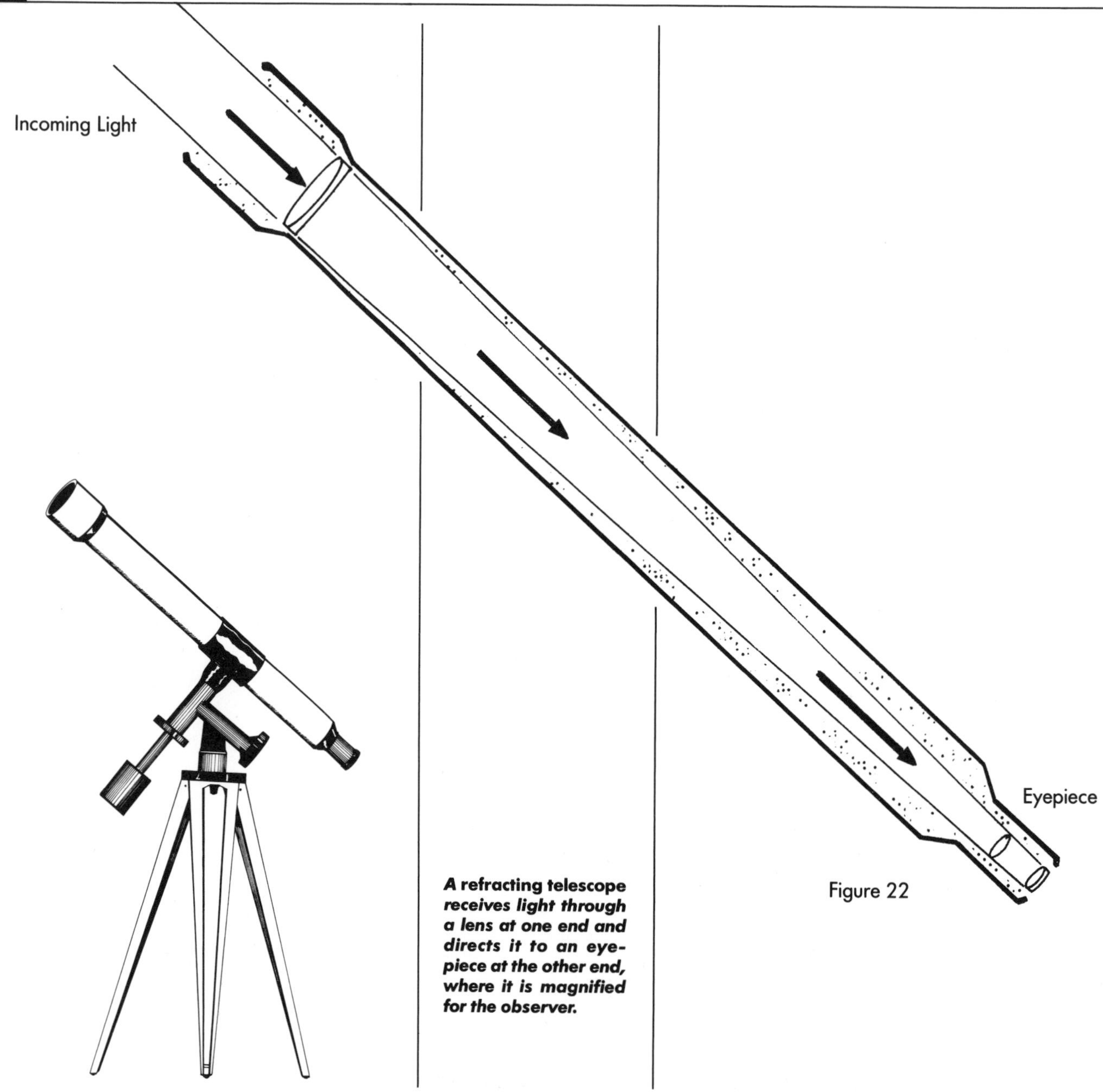

Incoming Light

Eyepiece

Figure 22

A refracting telescope receives light through a lens at one end and directs it to an eyepiece at the other end, where it is magnified for the observer.

period in the Moon's history. Besides craters, you'll see mountains, even entire mountain ranges, as well as cracks and valleys. Some of the lunar valleys (called rills) appear, remarkably, as if cut by an ancient river. More likely, they are the result of early molten material, because scientists believe there never has been any water on the Moon. The searing heat of the Sun (in the lit areas of the Moon), and the freezing cold of space (in the shadows), make the Moon's temperatures inhospitable for water.

Now that you have taken a good look at many of the Moon's interesting features, try drawing some of your favorite ones. Choose an area you know you will be able to find again, such as the detail around an obvious crater, then take a look on another night to see how much the more intricate features seem to have changed. Be sure, as always, to note the exact time and viewing conditions of your observations. As with all scientific procedures, repeat your observations. Take a look at the same area of the Moon at the same amount of time after new moon during the next lunar cycle. Also, try to see more detail by looking at the feature you have chosen during the waning phases (after full moon). The Sun and shadows will highlight the feature from another angle, allowing you to see new details. Make a new drawing, or improve on your first one during your second observation session.

The Moon moves in its orbit enough in two days for changes to be apparent. Over a period of ten days, major changes will be visible.

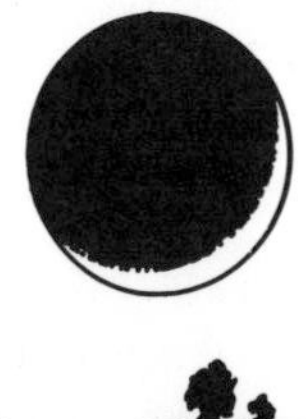

Figure 23

4 MEASURING LUNAR MOUNTAINS AND CRATERS

LEVEL: Advanced

EQUIPMENT: Telescope (or spotting scope with higher power), pencil, paper or Observation form, scientific calculator or personal computer, lunar map with latitude and longitude markings, and a current *Astronomical Almanac*

PROJECT: Estimate the diameters of lunar craters and calculate the height of lunar features by estimating the length of their shadows.

As you become more familiar with the Moon, you may want to know more about its features. This project will help you estimate their size.

Diameters of lunar craters are easiest to measure, and the skills you acquire in doing so will help you estimate the heights of lunar mountains. Because the Moon is a sphere, all features except those right at the center of the lunar disk will be "foreshortened"—more or less depending on their distance from the center of the Moon's disk. When an object is foreshortened, it appears compressed in one dimension; i.e., its width or length appears shorter than it actually is. If you tip a piece of notebook paper sideways and look at the long edge, you will see that its width seems to shrink. The trick to determining the actual size of a crater lies in knowing that most craters are nearly circular. Although foreshortening causes most craters to appear elliptical, this narrowing only affects a circle in one dimension. Hence, a crater's largest apparent dimension can be used to extrapolate its actual size. All you need is a good estimate of that dimension as a fraction of the apparent radius of the Moon. Once you have made that estimate, multiply your number by 1,080 to get the crater's radius in miles. For the diameter of the crater in kilometers, multiply by 1,738.

A close-up look at lunar features can show stark details that seem to change as shadows shift.

Drawings or photographs of the Moon can help in this process.

For example, make a circle to represent the entire lunar disk. Place a small oval in the proper place within the circle to represent the large crater Tycho, which displays bright rays from its rim across much of the lower part of the lunar surface. Be careful to make your drawing accurate in scale; your oval should be as close in size compared to your circle as the crater seems to be when compared to the entire Moon. When measured along its largest dimension, this representation of Tycho should appear to be about $\frac{1}{17}$ that of the lunar radius. Because the actual radius of the moon is about 1,080 miles, multiply your fraction, $\frac{1}{17}$, by 1,080. The result in this case is 63.5 miles for the diameter of Tycho. The diameter of Tycho is actually about 56 miles, but your estimate is not bad for a first attempt, using only a piece of paper, a pencil, and your eye to estimate the diameter of a crater 250,000 miles away!

Once you have some practice estimating the diameters of craters, you will be ready to move on to a more difficult set of observations—estimating lunar shadow lengths and calculating the heights of lunar mountains.

Begin by choosing a mountain you would like to measure. Then find a "reference" crater near the mountain that you can use as a measuring standard. You will also need to find out from your lunar map the selenographic (lunar) latitude and longitude of the feature you want to measure. You can obtain the size of the reference crater from your own estimates, but many books on the Moon contain more accurate numbers. If you are estimating the crater width yourself, remember to account for foreshortening (always use the largest apparent dimension to guarantee the most accurate reading of the crater's true diameter). Now compare the length of the shadow your mountain casts with the diameter of the reference crater and record

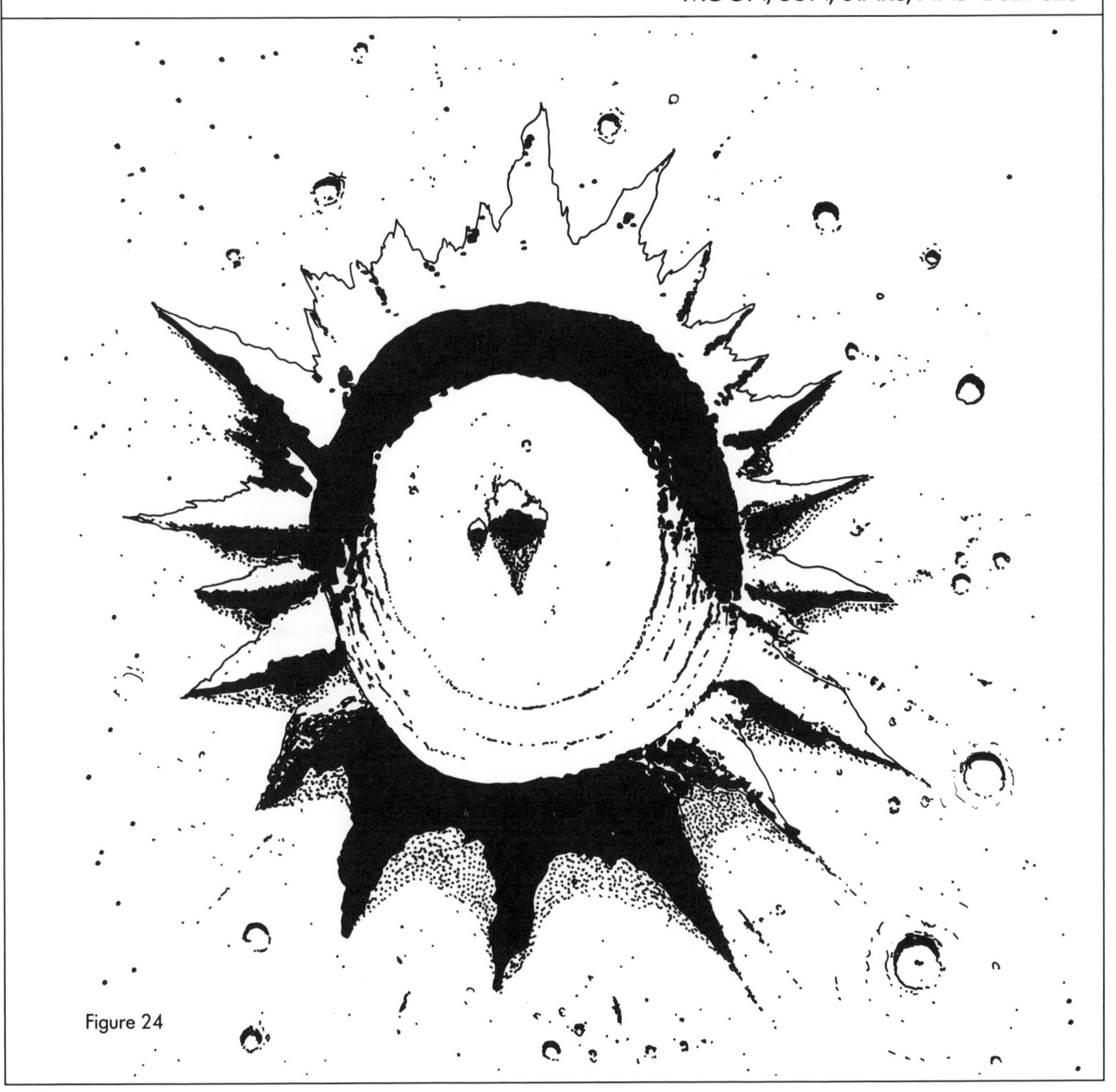

Figure 24

the exact time of your observation (at least to the nearest minute). Use the Universal time signals from channel WWV if possible.

Once you have made your observations, find the selenographic (lunar) longitude and latitude of the feature you are measuring by consulting the latitude/longitude grid of your lunar map. A lunar atlas would provide more accurate latitude and longitude figures, but you will need to perform some calculations to translate them into useful numbers. All the formulas you will need for this project follow, and the calculations will not be difficult for anyone who has used a calculator or computer to do trigonometry.

If you plan to use a lunar atlas, the best listing of the coordinates of lunar features can be found in the 1935 *IAU Lunar Atlas*. When using this atlas, use the coordinates *n* and e. To obtain selenographic latitude (B), use the formula $B = \sin^{-1} n$. For longitude (L), use $L = \sin^{-1}(e \sec B)$. If these letters and calculations look daunting, don't worry. The formulas that follow will enable you to let your calculator or computer do all the work. For your reference, $\sin^{-1}(x)$ means "the number whose sine is x." Most scientific calculators merely require you to put the number(s) in parentheses and then execute the appropriate function. Usually, "SIN^{-1}" will appear on the keyboard. If not, there will be a key marked "INV," for "inverse," which you must press before hitting "SIN." "Sec" stands for "secant"; "cos" for "cosine"; and "cosec" for "cosecant," which you can derive from Table 3, as many scientific calculators do not have this trigonometric function.

TABLE 3
Scientific Calculator Functions

To derive	Use calculator functions
Secant (A)	1 ÷ Cosine (A)
Cosecant (A)	1 ÷ Sine (A)

Once you have determined these numbers, find the Sun's selenographic latitude (B") and colongitude (C) in the Astronomical Almanac. These are the standard units used to define the point on the moon's surface at which the Sun would be directly overhead at a given time. Convert your time to a Universal time (see chart opposite) and express minutes as a decimal number to help make any interpolations necessary in using the almanac tables. You will also need to look up the Earth's selenographic longitude (L′) and latitude (B′). Write all these values down for the time of your observation, then calculate the height of the Sun (A) over your mountain using the formula

$$A = \sin^{-1}[(\sin B)(\sin B'') + (\cos B)(\cos B'')\sin(C + L)].$$

Next, calculate the Earth/Moon/Sun angle (F) with the formula

$$F = \cos^{-1}[(\sin B')(\sin B'') + (\cos B')(\cos B'')\sin(L' + C)].$$

Finally, calculate the height of your mountain (H) as a fraction of the lunar radius with

$$H = D(\sin A)(\text{cosec } F) - \tfrac{1}{2}(D^2)(\text{cosec}^2 F)(\cos^2 A),$$

where D is the shadow length you estimated as a fraction of the Moon's radius.

As you carry out each of these calculations, write down (or store in your calculator's memory) each result to use in the next equation. The final step is, again, to multiply the height (H) you have calculated by 1,080 to get the height in miles, or by 1,738 for the height in kilometers.

This may seem complex, but with a good scientific calculator or personal computer you can readily make all these calculations. Once you have learned how to use a scientific calculator, you may find that the most difficult part is deciding which lunar mountain to measure.

UNIVERSAL TIME CONVERSION CHART

Universal Time is the time standard used around the world. To calculate your local time from Universal Time, subtract the following number of hours for your time zone:

Eastern Standard Time: subtract 5 hours
Central Standard Time: subtract 6 hours
Mountain Standard Time: subtract 7 hours
Pacific Standard Time: subtract 8 hours
Most of Alaska: subtract 9 hours
Hawaii: subtract 10 hours

For daylight saving time, which usually runs from the first Sunday in April through the last Sunday in October, take away one hour from the numbers above (4, 5, 6, 7, 8 and 9).

Figure 25

A scientific calculator makes astronomical calculations easy.

5 ECLIPSES

LEVEL: Intermediate

EQUIPMENT: Star maps, sighting stick, pencil, and observations from Project 1, Phases of the Moon

PROJECT: Plot the Moon's position against the stars and the ecliptic; determine likely months for eclipses and compare your results with the eclipse chart. Observe eclipses where possible

Before you begin this project, please read this cautionary paragraph carefully. Eclipses of the Sun can be among the most spectacular sights in the sky, especially if you are lucky enough to be on the narrow band of Earth where one can see a total solar eclipse. But, because of its brightness, the Sun is highly dangerous to observe. NEVER look at the Sun directly, either with the naked eye or through any optical aid, at any time, even when its disk is almost completely obscured by the Moon. You can safely watch the moments of total eclipse, but you MUST take great care to look away as soon as the glare begins to increase and the Moon's disk shifts enough to reveal even the slightest bit of the solar disk. Further, NEVER look at a solar eclipse, even during totality, through binoculars, telescopes, or any other optical instrument that is not properly shielded. Doing so can cause permanent, serious eye damage. Figure 26 shows one safe way to view a solar eclipse. Lunar eclipses, on the other hand, are always safe to view.

Eclipses result from certain alignments of the Sun, Earth and Moon. In a solar eclipse, the Moon blocks the Sun from the view of people

Projecting the Sun's reflection from a small mirror in an envelope with a hole in it onto a white board is one safe way to view a solar eclipse.

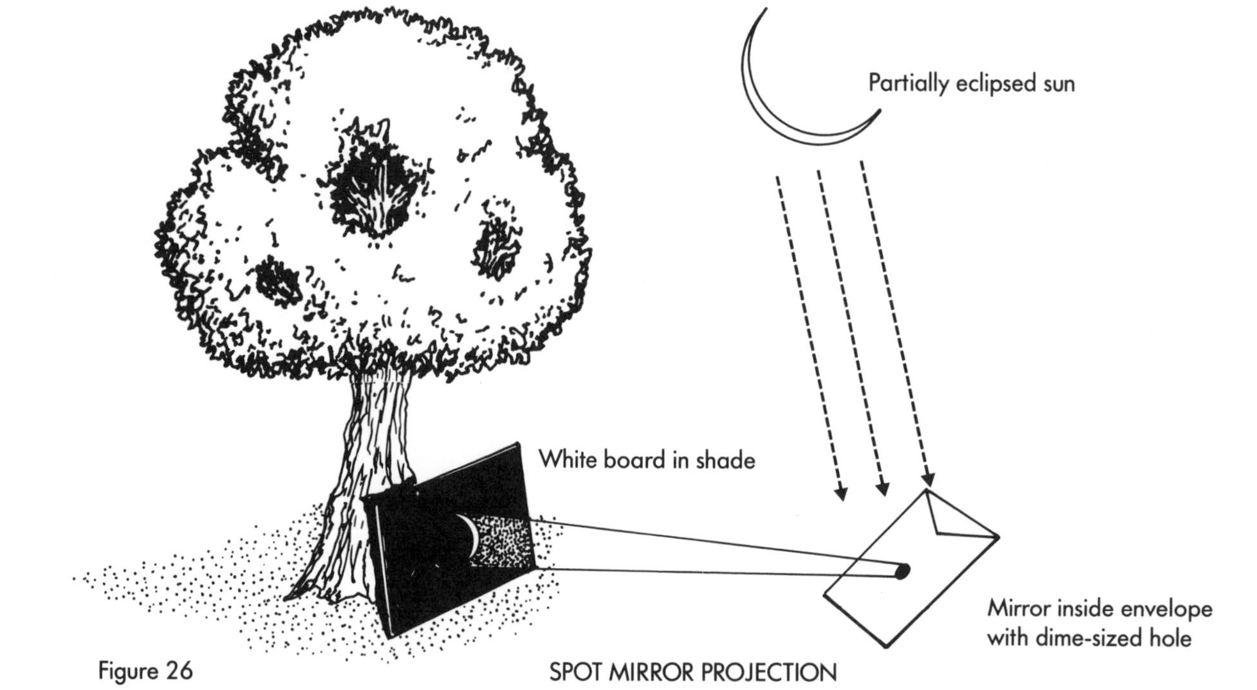

Figure 26 SPOT MIRROR PROJECTION

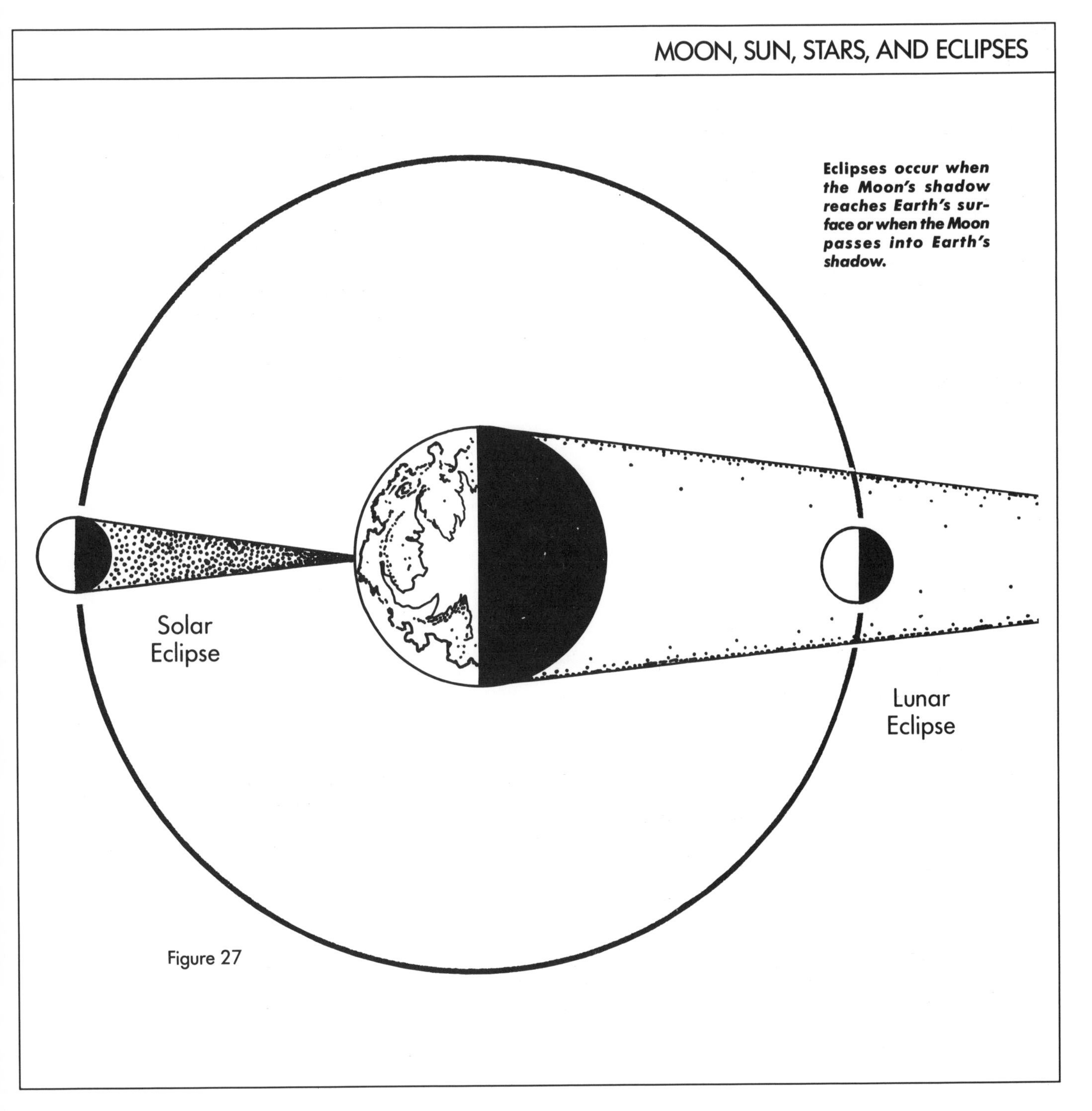
Eclipses occur when the Moon's shadow reaches Earth's surface or when the Moon passes into Earth's shadow.
Solar Eclipse
Lunar Eclipse

Figure 27

on a certain band of Earth; in a lunar eclipse, the Moon slips into the shadow of the Earth. Depending on the exact configuration, people on different parts of the Earth will see different eclipses in varying degrees, or possibly see none at all. In the case of total and "annular" solar eclipses, there is only a very thin path, perhaps only a few miles wide, where the full effect of the eclipse is visible.

Even on the path where the full eclipse is visible, totality, or maximum eclipse, lasts only a few minutes. Still, people are so taken with solar eclipses that they travel to distant parts of the world to see them, often planning their trips years in advance.

Lunar eclipses, on the other hand, are usually visible over a wide area of our planet. A total lunar eclipse lasts for hours, with the eclipse lasting for much of the night if conditions are right. From any location where you can see the Moon above the horizon, you can see the eclipse if the weather cooperates. Though not as dramatic as solar eclipses, lunar eclipses are beautiful in their own way. Tables 4 and 5 contain the dates of the eclipses that will be seen somewhere on Earth during the 1990s. You can compare this table with the estimates from your observations in Project 1.

TABLE 4
Lunar Eclipses

Date	Type	Beginning	End	Extent
June 4, 1993	t	6:11 A.M.	9:50 A.M.	157%
Visibility: Australia, Far East. Partial: eastern Asia, western North and South America				
Nov. 28–29, 1993	t	11:40 P.M.	3:12 A.M.	109%
Visibility: North and South America. Partial: Europe, western Africa, Japan				
May 24, 1994	p	9:38 P.M.	11:23 P.M.	25%
Visibility: USA, Canada, South America, parts of Europe, Africa				
Apr. 15, 1995	p	6:41 A.M.	7:55 A.M.	12%
Visibility: Australia, New Zealand, Pacific, Japan, Indonesia				
Apr. 3, 1996	t	5:21 P.M.	8:59 P.M.	138%
Visibility: Europe, Africa, western South America. Partial: North America (except western), central Asia				
Sep. 26, 1996	t	8:12 P.M.	11:36 P.M.	124%
Visibility: North and South America, western Europe, western Africa				
Mar. 23–24, 1997	p	9:58 P.M.	1:22 A.M.	92%
Visibility: Canada, USA, South America, Europe, parts of Africa				
Sep. 16, 1997	t	12:08 P.M.	3:25 P.M.	120%
Visibility: Asia, eastern Europe, eastern Africa, Australia				
July 28, 1999	p	5:22 A.M.	7:45 A.M.	40%
Visibility: Australia, New Zealand, east Asia, Pacific				
Jan. 20–21, 2000	t	10:01 P.M.	1:26 A.M.	133%
Visibility: North and South America, extreme western Europe, extreme western Africa				
July 16, 2000	t	6:57 A.M.	10:54 A.M.	177%
Visibility: eastern Asia, Australia				

p = partial; t = total

TABLE 5
Solar Eclipses

Date	Type	Peak of eclipse	Duration of peak obscurity	Maximum extent of solar disk obscured
May 21, 1993	p	9:19 A.M.	—	74%
Visibility: Canada, USA, Greenland, northern Europe, northern Russia				
Nov. 13, 1993	p	4:45 P.M.	—	93%
Visibility: Antarctica, New Zealand				
May 10, 1994	a	12:11 P.M.	6 min, 14 sec	94%
Nov. 3, 1994	t	7:39 A.M.	4 min, 23 sec	106%
Apr. 29, 1995	a	12:32 P.M.	6 min, 37 sec	95%
Visibility: Ecuador, Peru, Colombia, Brazil; partial in southern South America, Central America, Caribbean, southern Mexico				
Oct. 23–24, 1995	t	11:32 P.M.	2 min, 9 sec	102%
Apr. 17, 1996	p	5:37 P.M.	—	88%
Visibility: New Zealand, southern Pacific				
Oct. 12, 1996	p	9:02 A.M.	—	76%
Visibility: Europe, northern Africa, North Atlantic				
Mar. 8, 1997	t	8:24 P.M.	2 min, 50 sec	104%
Sep.1, 1997	p	7:04 P.M.	—	90%
Visibility: Australia, New Zealand, Pacific				
Feb. 26, 1998	t	12:28 P.M.	4 min, 9 sec	104%
Aug. 21, 1998	a	9:06 P.M.	3 min, 14 sec	97%
Feb. 16, 1999	a	1:34 A.M.	0 min, 40 sec	99%
Visibility: Indian Ocean, western and northern Australia; partial in southern Africa, Australia, Indonesia, Antarctica				
Aug. 11, 1999	t	6:03 P.M.	2 min, 23 sec	103%
Feb. 5, 2000	p	7:49 A.M.	—	58%
Visibility: Antarctica				
July 1, 2000	p	2:33 P.M.	—	48%
Visibility: South Pacific				
July 30, 2000	p	9:13 P.M.	—	60%
Visibility: the Arctic, Alaska, northern Russia				
Dec. 25, 2000	p	12:35 P.M.	—	72%
Visibility: Canada, USA, Mexico, Caribbean				

p = partial; t = total; a = annular; min = minutes, sec = seconds

SOLAR ECLIPSES

You will not have as many viewing opportunities for solar eclipses as for lunar ones, but if you do have a chance to be in the path of a solar eclipse, observe it cautiously and carefully. Use the projection technique shown in Figure 26, or place the proper filter on the front end of your telescope (Figure 28). Inexpensive "eclipse glasses" are popular at these events, and can be used safely, but make absolutely certain that they are made from quality material and have no scratches or holes in them. If they do have scratches or holes, DO NOT USE THEM. If you have a telescope but no filter for the front, watch the eclipse by projecting it on a card held a foot or more away from the eyepiece. Or, build a Sun viewer as described in the Tools of the Trade section (pages 28–29). Also, see the Sources list (page 120) for information on the purchase of a Sun viewer. DO NOT use filters that attach to the eyepiece or allow anyone to touch or even get near the eyepiece when the telescope is pointed at the Sun without filtering over the front end. The Sun's heat is so intense that it can crack filters, and its light is so bright that by the time a person looks away, permanent and serious eye damage will have already occurred.

For centuries, astronomers have used the projection technique to observe solar eclipses safely.

When using a filter to view a solar eclipse, check it carefully to make sure it has no scratches or holes, and always use one that attaches to the front end, rather than the eyepiece, of your telescope.

When observing a solar eclipse, you can make some of the same observations as you did with lunar eclipses. What time do you see the dark disk of the Moon starting to encroach upon the Sun (first contact)? When and how complete is the maximum eclipse, from your location? When does last contact—the moment

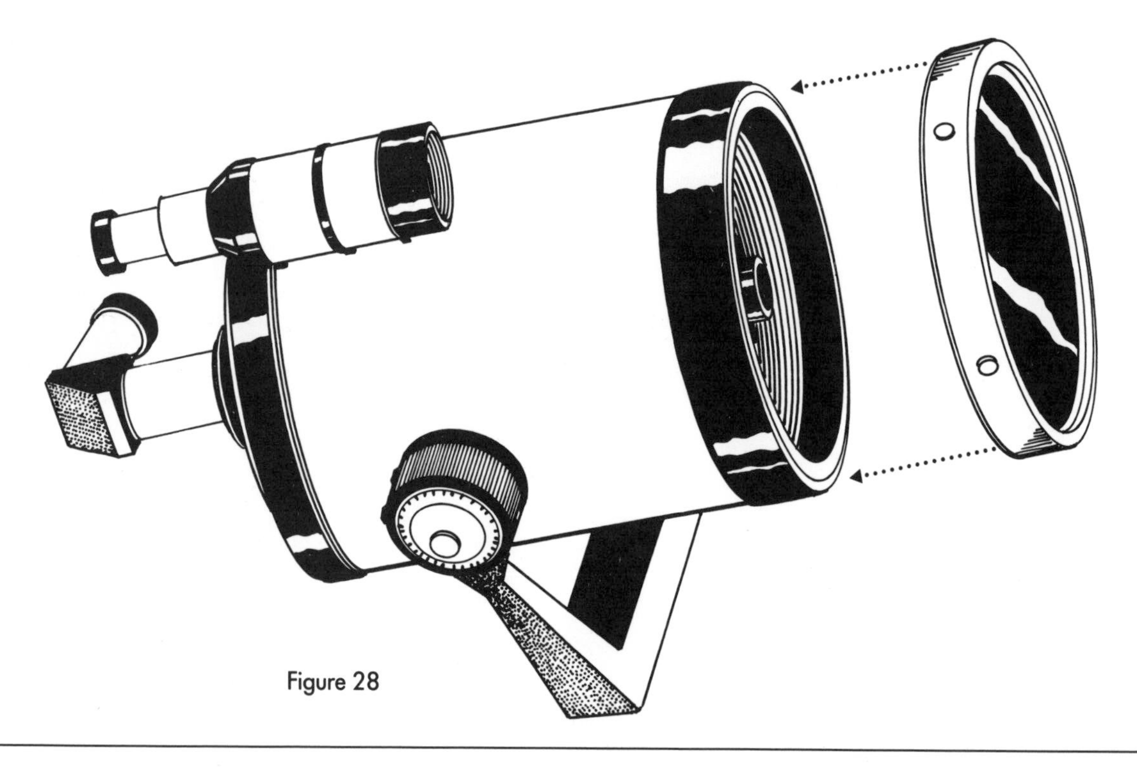

Figure 28

Figure 29

when the last bit of the Moon leaves the Sun's face—occur? Do any sunspots (darker patches on the Sun's surface) become visible during the eclipse?

If you can make the opportunity (or you are lucky to just be in the right place) to see a total solar eclipse, watch the effects during totality and record, photograph, or draw them. These include the shape of the *corona* (the outer atmosphere of the Sun), *prominences* (flames that shoot out from the solar surface), *Bailey's beads* (bright spots that shine through low points at the lunar limb, or edge), and many others. You might even be able to spot comets that are near the Sun or bright planets too closely aligned to the Sun at the time to see in the night sky. Recording any or all of these events can add to the data astronomers need to collect about eclipses as well as to your understanding of the phenomena.

The area on Earth from which you can see a total solar eclipse is very narrow.

If you did Project 1 completely, you should have what you need to get started here. If not, try using the sighting stick to determine the positions of the Moon for a week or more. One of the last suggestions in Project 1 was to plot the Moon's position on copies of the star maps and see how its path differs from the ecliptic (i.e., the Sun's apparent path through the sky). Use these observations to estimate the approximate location of the Moon's path as compared to the

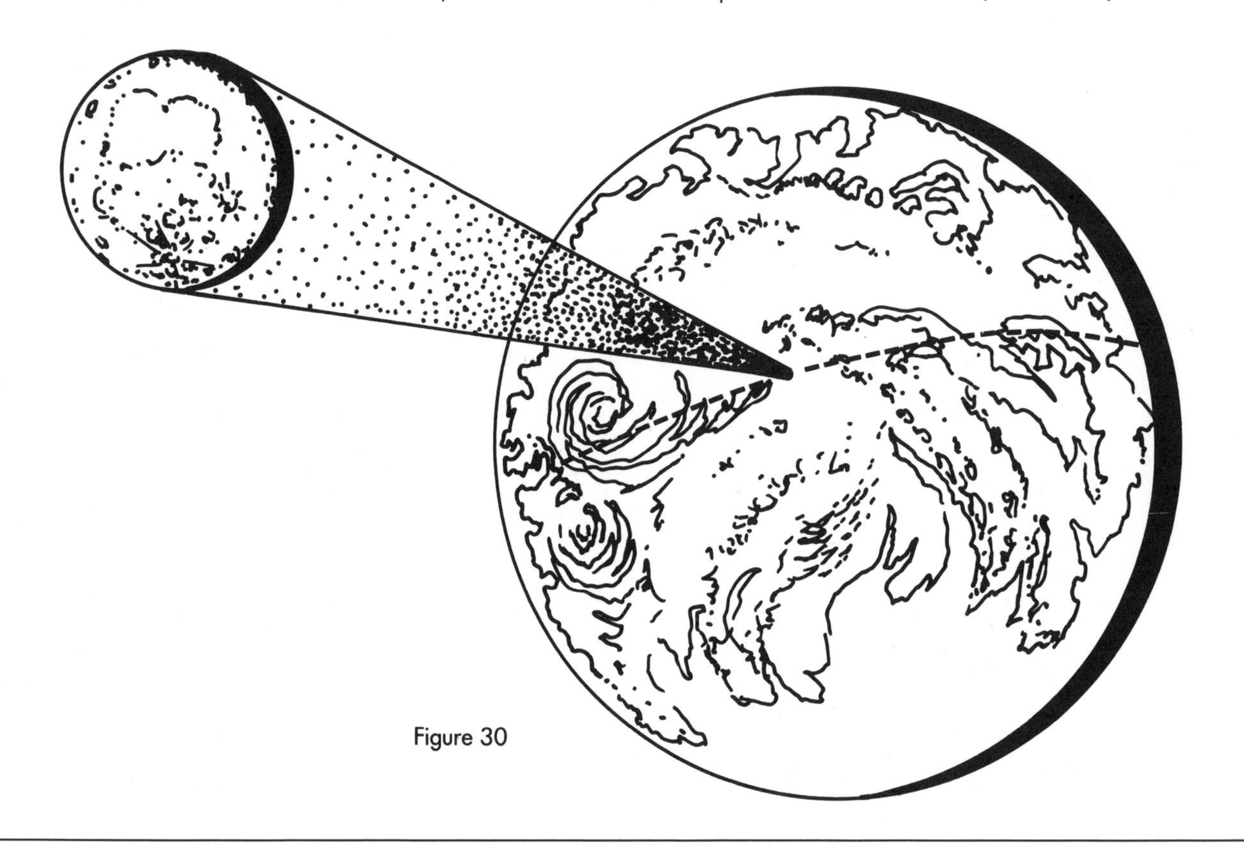

Figure 30

ecliptic. Be careful to determine the proper path, which shifts from lunar cycle to lunar cycle. Try to determine when the path might cross the ecliptic. You can use your observations to extrapolate other, later times when the Moon will cross, which it does twice during each lunar cycle. If it crosses exactly at new moon or full moon, there will be a solar or lunar eclipse, respectively. Although the Moon goes around the Earth in about 27.3 days, because the Earth orbits the Sun simultaneously, it takes about 29.5 days for a particular phase to return. You can use this timing, for a few months, to predict where the Moon will be, and by using the markings along the ecliptic on the star maps you can also find out where the Sun will be during the times in question.

Using your observations, try to estimate when eclipses will occur in the near future. You will discover that solar and lunar eclipses usually come in pairs, about two weeks apart, and that there are normally no more than two pairs of eclipses anywhere in the world during a given calendar year. Try to find out, through your nearest planetarium, astronomy club, or references such as the Royal Astronomical Society of Canada's *Observer's Handbook* or *Astronomy* magazine, when and where eclipses will actually be visible. Try to observe them. If possible, observe some eclipses and check the accuracy of your predictions.

LUNAR ECLIPSES

For any location on Earth, at least one partial lunar eclipse should be visible each year. You may be surprised, in fact, to see some day what at first glance looks like a crescent moon rising in the east at sunset but is actually a lunar eclipse in progress. You will know it is an eclipse because a crescent moon normally appears in the western sky at sunset. More likely, however, if you've been watching for eclipses, you will have an opportunity over the next few years, without travelling beyond your own neighborhood, to see the Moon slip at least partway into the shadow of the Earth at full moon. There is no apparent change at first. The lighter inner shadow does not visibly change the brightness of the Moon. As the Moon passes farther into the Earth's shadow, however, one edge will begin to appear darker. Try timing when you notice this first change.

Unless the Moon enters the inner shadow (umbra) of the Earth, an eclipse is called *penumbral*. Some penumbral eclipses go completely unnoticed by the casual observer. Once the Moon is in the Earth's inner shadow, however, a very distinct dark patch with a curved edge becomes apparent. If the Moon is never completely covered by the Earth's shadow, the eclipse is partial. Time the point at which the shadow appears to start retreating from the lunar disk. Can you see any coloring in the dark area?

When the Moon moves completely into the Earth's umbra, a total lunar eclipse occurs. These make a beautiful sight because the Moon is bathed in reddish light refracted around the Earth by our atmosphere. The exact coloring depends on how deeply the Moon passes into the inner shadow, what atmospheric conditions (such as air pollution, volcanic ash, and even weather patterns) prevail at the Earth's terminator, and local visibility conditions at the time. During some total lunar eclipses, the Moon's disk is white on one edge, fading to bright orange at the other; sometimes the entire disk will be a very deep brownish color. See if you can determine when all the white disappears, and record your overall impression of the coloring. Try drawing the Moon during eclipse. Time the point at which the Moon seems to lighten up again. All of these are interesting observations to make during lunar eclipses.

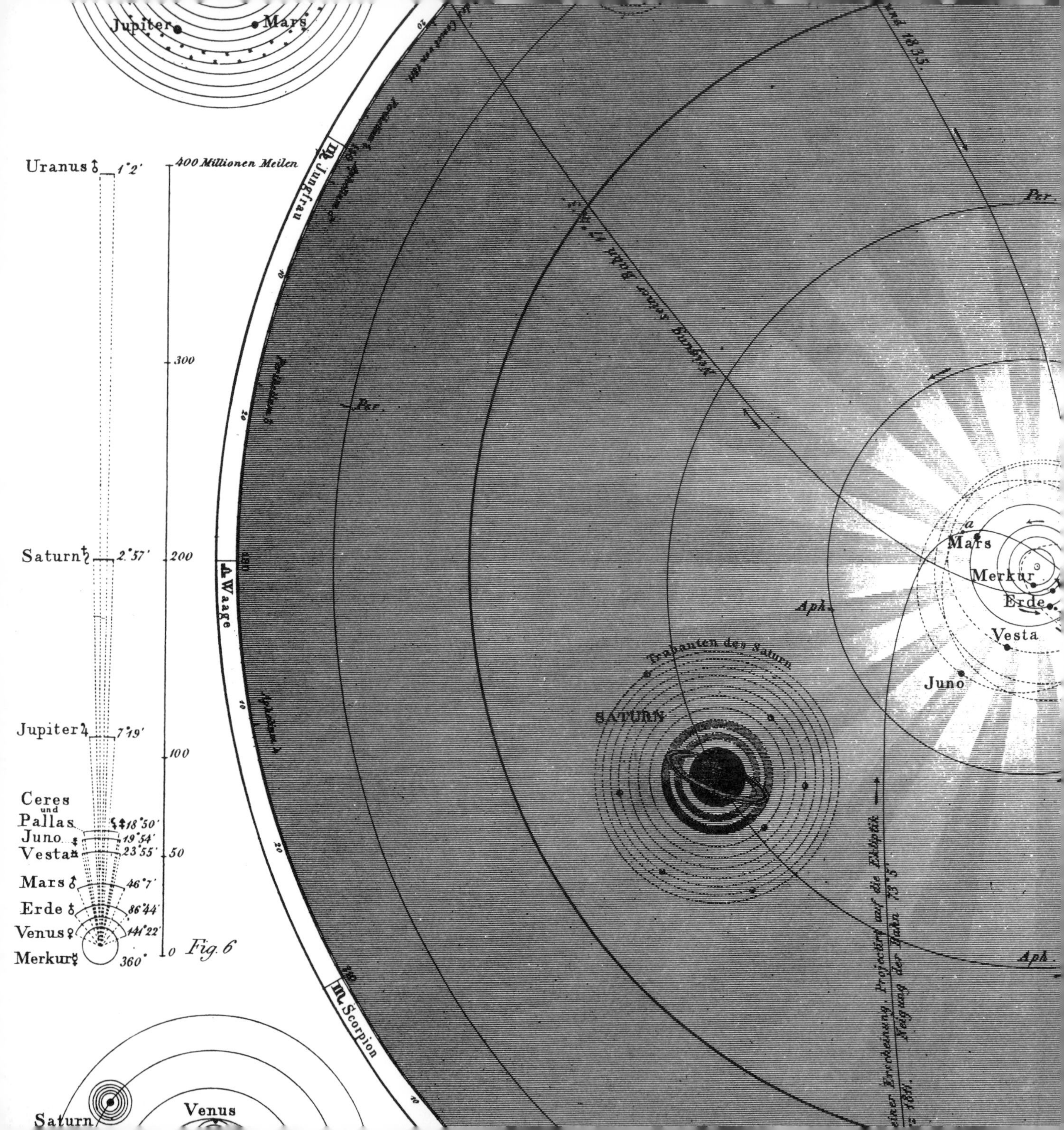

Jupiter
Mars
Uranus ♅ 1°2′
400 Millionen Meilen
300
Saturn ♄ 2°57′
200
Jupiter ♃ 7°19′
100
Ceres und Pallas 18°50′
Juno 19°54′
Vesta 23°55′
50
Mars ♂ 46°7′
Erde 86°44′
Venus ♀ 141°22′
Merkur ☿ 360°
0
Fig. 6
Jungfrau
Waage
Scorpion
Par.
Aph.
Trabanten des Saturn
SATURN
Mars
Merkur
Erde
Vesta
Juno
Saturn
Venus

SECTION 3
THE SOLAR SYSTEM

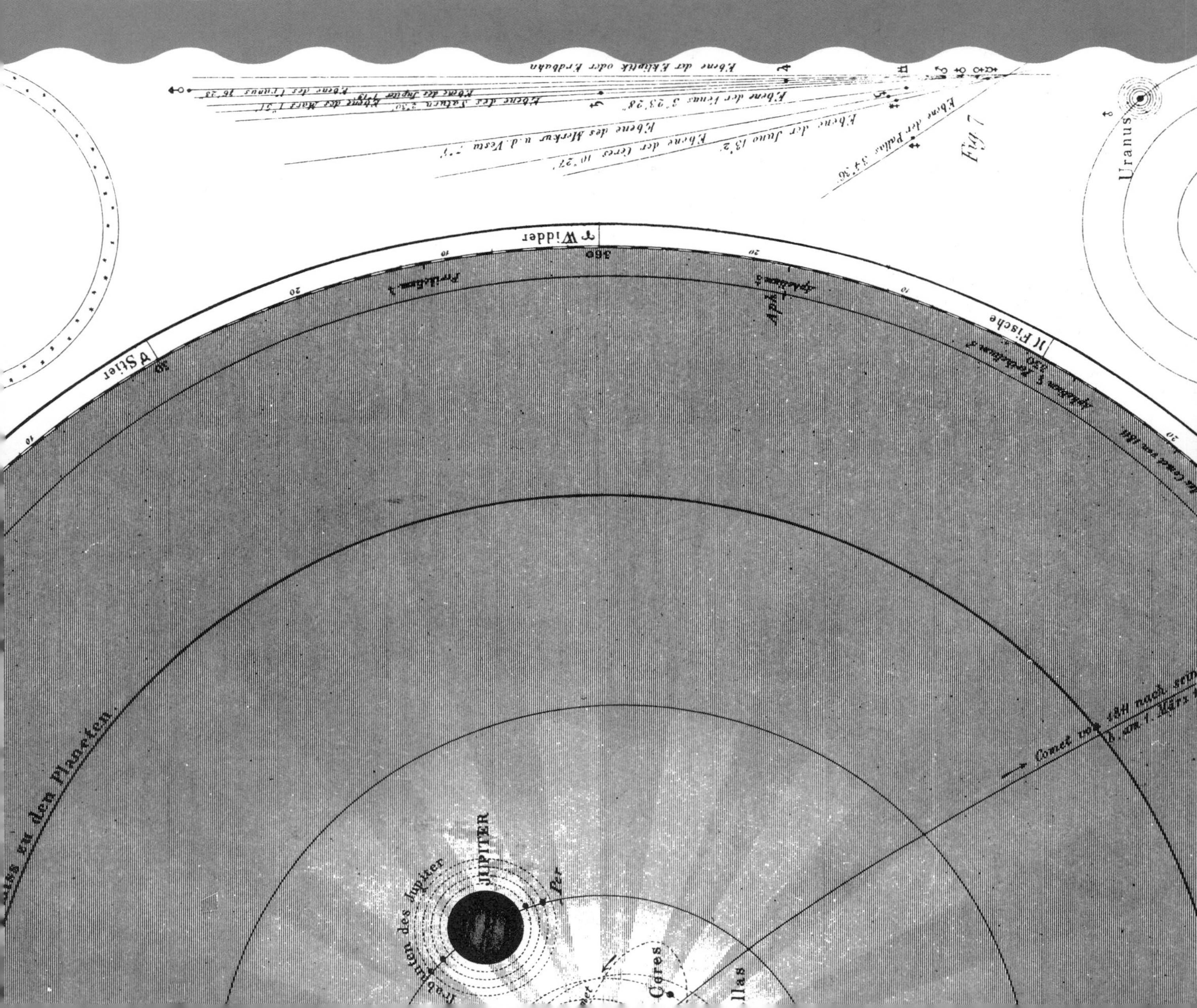

6 RELATIVE SIZES OF THE PLANETS

LEVEL: Basic

EQUIPMENT: Common objects of various sizes—ideally, a basketball, a round balloon, beads, other spherical objects, and pins; two-dimensional objects (cut out of or marked on paper or poster board) can be substituted for these if necessary

PROJECT: Compare the relative sizes of different objects to get an idea of how the planets compare in size

On a clear, dark, moonless night, nearly three thousand stars can be seen with the unaided eye. Each of these stars is a thermonuclear furnace burning hydrogen at its center. This burning generates unimaginable amounts of light and heat, causing the star to shine so brightly that we can see it even at so-called astronomical distances. These distances are truly immense. The light from the nearest star (besides the Sun) takes more than four years to reach the Earth, and light travels at more than 186,000 miles, (299,000 km) per second! Even many of the fainter stars we see in the night sky would far outshine our Sun if put in its place. The Sun is just a medium-sized, medium-bright star, but our relative closeness to it (about 93 million miles, or 150 million km) gives us all the light and heat we need to live on Earth, plus enough energy to provide for our electrical needs for centuries to come (if we can harness it).

The planets, on the other hand, are tiny in comparison with stars. They reflect sunlight, rather than emitting their own light. All the objects in our solar system, however, are much closer to us than any of the stars (except, of

The nine planets, including Earth, that comprise our solar system, orbit the Sun.

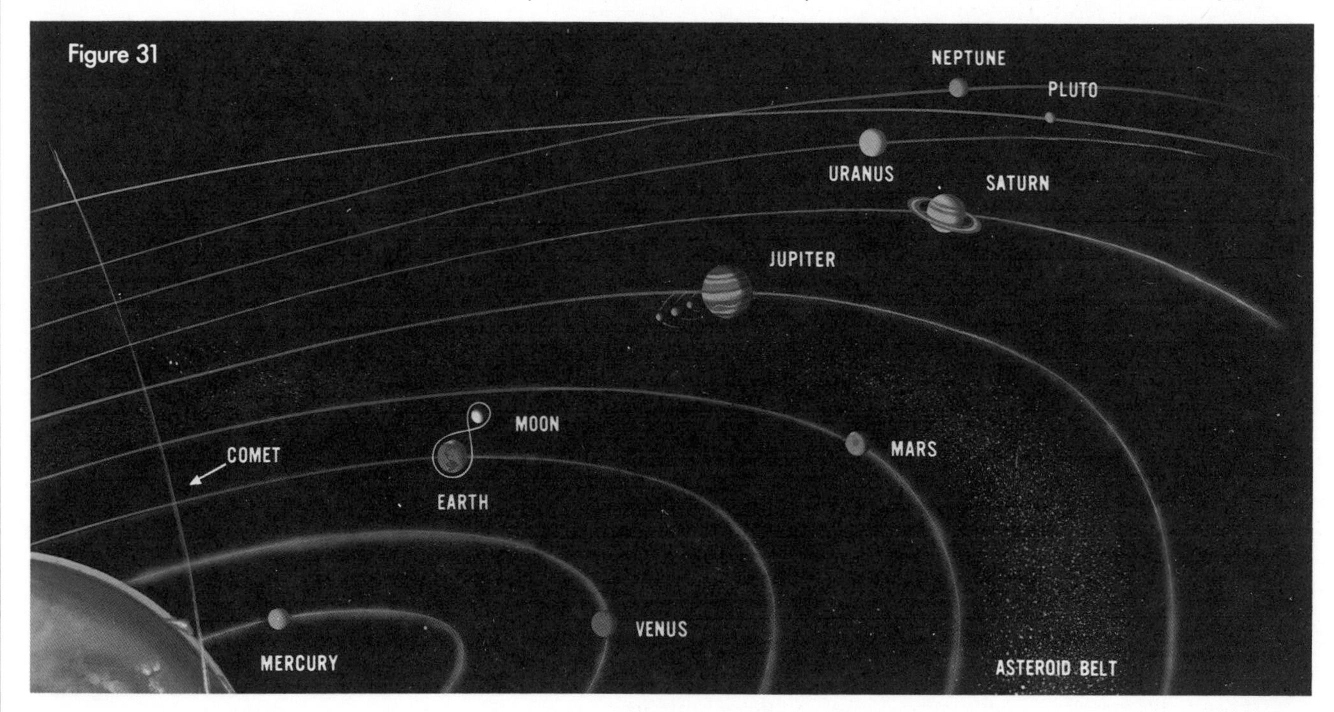

course, the Sun). This means that they are much easier to observe in detail. It does, however, take a fairly large telescope to see any details on the outer planets of Uranus and Neptune, and the outermost planet, Pluto, at a distance of 3.6 billion miles (5.8 billion km) is just a smudge in the best telescopes. The bright planets, Saturn, Jupiter, Mars and Venus, and, with a little more difficulty, Mercury, can easily be observed in detail.

The appearance of the planets differs from that of the stars in several respects. The bright planets—and Uranus and Neptune, too—are always found near the ecliptic. (Pluto is a bit of an exception because of its unusual orbit.) The planets seem to move against the background stars in unusual ways, because they and the Earth are all moving around the Sun. But the planets appear different from the stars in several other ways, too. Besides their greater brightnesses and their unique colors, the planets do not "twinkle" as the stars do. This, their brightness, and their motion against the background stars all make them easy to find in the sky.

Common objects can help you visualize the relative sizes of the Sun and the planets.

Figure 32

Basketball (SUN)

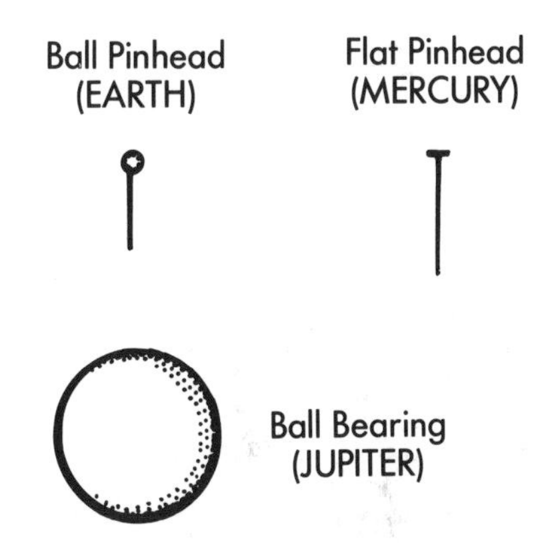

Project 6 and Project 7 are activities that the whole family can have fun with on a rainy night, indoors. But they will still teach you some fascinating facts about the solar system. Your most difficult task will be finding objects of the right size to use as your set of scale-model planets. If you're doing these projects with children, try to use a globe of the Earth as an example of a scale model of our planet. Once you have gathered together all the objects you need, compare the globe or even a map of the world to the pinhead you will use to scale all the planets and the Sun down to sizes that will fit into a room. Then compare the objects to get a sense of the sizes of the planets relative to each other.

6 RELATIVE SIZES OF THE PLANETS

Table 6 provides measurements of some common objects you can use to represent the Sun and planets. Compare these measurements, remembering that the Earth is actually 7,909 miles (12,756 km) in diameter. We have shrunk the Earth down nearly 550 million times to show it to scale with a basketball-size "Sun." Table 6 also contains the relative distances of the planets from our scale-model Sun. You can see that even at this small size, our pinhead Earth is 88 feet (27 m) out, and Pluto, on average would be more than two-thirds of a mile (a little more than a kilometer) from the "Sun." Even Mercury, about two-fifths of the distance from Sun to Earth, is about as far away from the scale-model Sun as the length of a small house.

TABLE 6
Solar System Dimensions
(relative to basketball-size Sun)

Body	Scaled-down size		Scale-model object	Relative distance from Sun*	
Sun	10.000 in	25.40 cm	Basketball	—	—
Mercury	0.035 in	0.89 mm	Straight pinhead	36 ft, 1.1 in	11.0 m
Venus	0.087 in	2.21 mm	Round pinhead	65 ft, 7.4 in	20.0 m
Earth	0.092 in	2.34 mm	Round pinhead	88 ft, 7.0 in	27.0 m
Mars	0.049 in	1.24 mm	Cold-capsule granule	137 ft, 9.5 in	42.0 m
Jupiter	1.025 in	26.04 mm	1-inch diameter bead	469 ft, 1.9 in	143.0 m
Saturn	0.862 in	21.89 mm	Large marble	839 ft, 10.7 in	256.0 m
Uranus	0.368 in	9.35 mm	Ball bearing	1,725 ft, 8.6 in	526.0 m
Neptune	0.349 in	8.86 mm	Ball bearing	2,700 ft, 1.6 in	823.0 m
Pluto	0.016 in	0.41 mm	Dot as tiny as period at end of this sentence.	3,553 ft, 1.7 in	1,083.0 m

*Distances from the Sun to each planet are average distances. Because of its eccentric orbit, in 1992 Pluto is closer to the Sun than Neptune and will remain so until 1999.

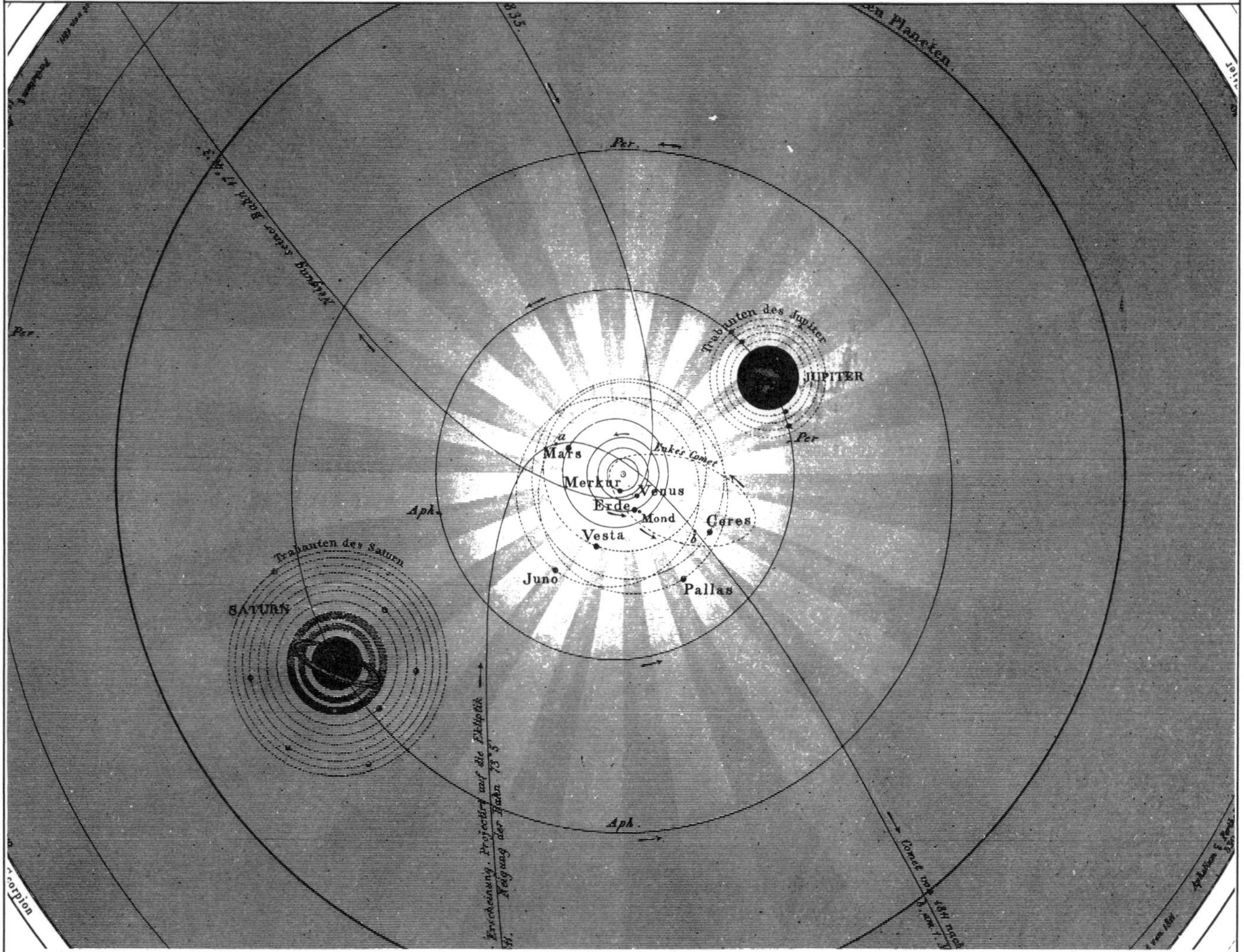

Figure 33

This centuries-old representation of the solar system depicts what were then thought to be the orbits of the planets.

7 SCALE OF THE SOLAR SYSTEM

LEVEL: Basic

EQUIPMENT: Ruler, yardstick or tape measure, paper, rope or yarn, scissors, index cards

PROJECT: Measure out, at a reduced scale, the distances between the Sun and each of the planets

You need not carry out Project 6 to undertake this one, but they are both more interesting when done together. Project 6 shows us how the planets compare to each other in size and distance from each other. If the Sun, which has an actual diameter of about 865,000 miles, or 1,392,000 km (see Table 7), is represented by a basketball (10 inches/25.4 cm in diameter), then Earth is the size of a round pinhead less than one-tenth of an inch across, and 88 feet (27 m) away from the "Sun." In Project 7 we need to scale things down even more so that the entire solar system can "fit" inside a small house. Unfortunately, this makes the planets too small to see (simply use dots to represent them), with the exception, perhaps, of mighty Jupiter, which would be a tiny dot one-hundredth of an inch across.

Measure lengths of string or yarn, using Table 8, to indicate the relative distances of the planets.

If you make the Sun about an inch (2.5 cm) in diameter, however, the Earth disappears and Jupiter is the size of the pinhead we previously used for the Earth. At this scale, Pluto is still 355 feet (108 m) from the Sun but now the Earth is just 9 feet (3 m) away.

Scale things down again, according to Table 8, and measure out some rope (Figure 34), or, better yet, different colored strings or yarns whose lengths match the distances shown for each planet in Table 8 Change colors or use a marker as you go out to show each planet distance clearly. Measure out 4.2 inches (.11 m) of red for the distance to Mercury, then add 3.6 inches (.09 m) of another color (to get a total of 7.8 inches/.20 m for the distance of Venus, and so on). Have someone hold one end of the string and move outward while someone else unravels it, taking note of the distances. In this final example, our pinhead Sun and its solar system are just under 550,000,000,000,000 (550 trillion) times smaller than in reality!

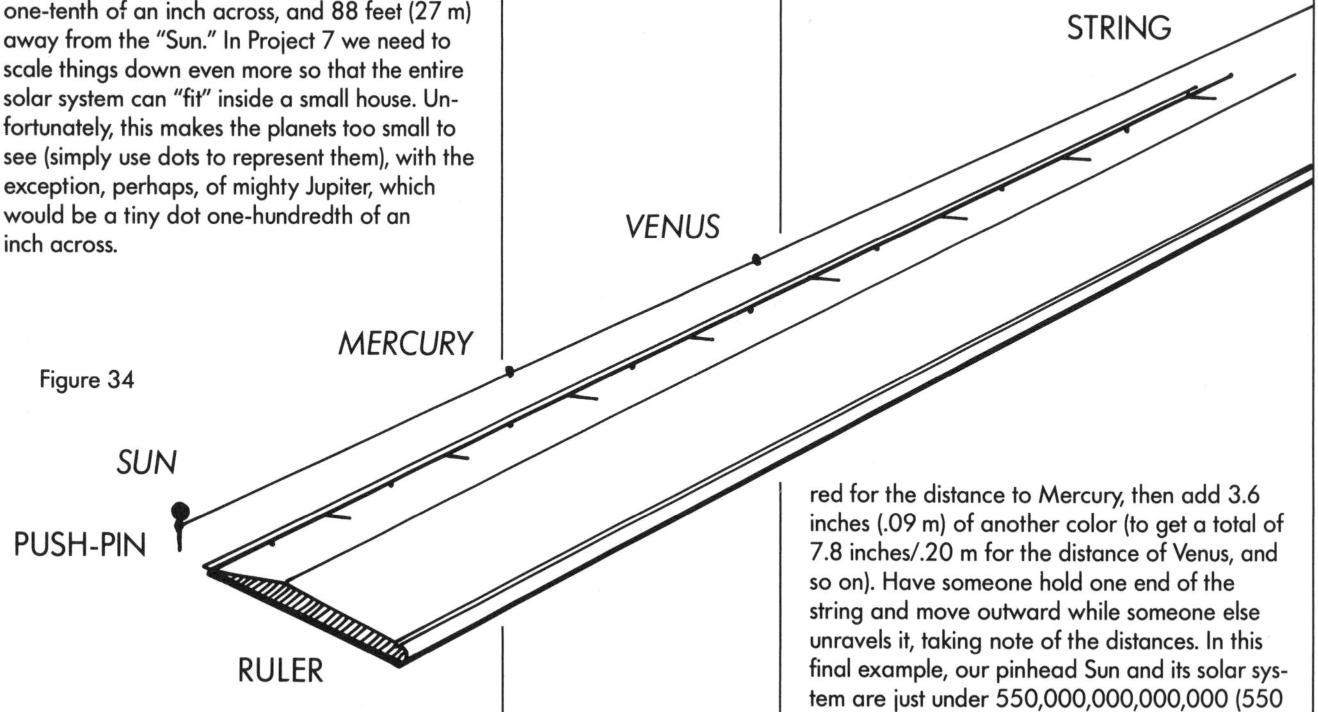

Figure 34

TABLE 7
Solar System Statistics

Solar system body	Average distance from Sun		Equatorial diameter		Period of revolution	Period of rotation
	in millions of km	Astronomical units	Equatorial Diameter in km	Compare distance to Earth	Earth years	Earth days
Sun	—	—	1,392,000	109.13	—	25–35*
Mercury	57.9	0.387	4,878	0.38	0.38	58.646
Venus	108.2	0.723	12,104	0.95	0.85	243.017
Earth	149.6	1.000	12,756	1.00	1.00	0.9973
Mars	227.9	1.524	6,787	0.53	1.83	1.0260
Jupiter	778.4	5.203	142,980	11.21	11.86	0.4101
Saturn	1424.5	9.522	120,540	9.45	29.46	0.4440
Uranus	2872.4	19.201	51,120	4.01	84.01	0.718
Neptune	4499.0	30.074	49,530	3.88	164.79	0.671
Pluto	5942.8	39.725	2,300	0.18	247.69	6.3867

*Depending on latitude.

TABLE 8
Solar System Dimensions
(relative to pinhead-size Sun)

Body	Scaled down		Relative distance from Sun*	
Sun	0.10 in (round pinhead)	2.5 mm	—	
Mercury	(microscopic)		4.2 in	0.11 m
Venus	(microscopic)		7.8 in	0.20 m
Earth	(microscopic)		10.8 in	0.27 m
Mars	(microscopic)		1 ft 4.4 in	0.42 m
Jupiter	(microscopic)		4 ft 8.2 in	1.43 m
Saturn	(microscopic)		8 ft 4.9 in	2.56 m
Uranus	(microscopic)		17 ft 2.9 in	5.26 m
Neptune	(microscopic)		27 ft 0.0 in	8.23 m
Pluto	(microscopic)		35 ft 6.4 in	10. 83 m

*Distances from the Sun to each planet are average distances. Because of its eccentric orbit, Pluto in 1992 is closer to the Sun than is Neptune and remains so until 1999.

Indianischer Vogel
Skorpion
Antares
Macula
Magellanica
Fliege
Paradiesvogel
Colur
Südl. Krone
Polarkreis
Mikroskop
Kranich
Südlicher Fisch
Fomalhaut

***Page 65:* The Moon.**

Page 66: Top; Astronaut next to lunar boulder.
Bottom left; Total solar eclipse.
Bottom right; Eclipse of the Sun by the Earth.

Page 67: Top; Mercury (colorized).
Bottom; Venus (colorized).

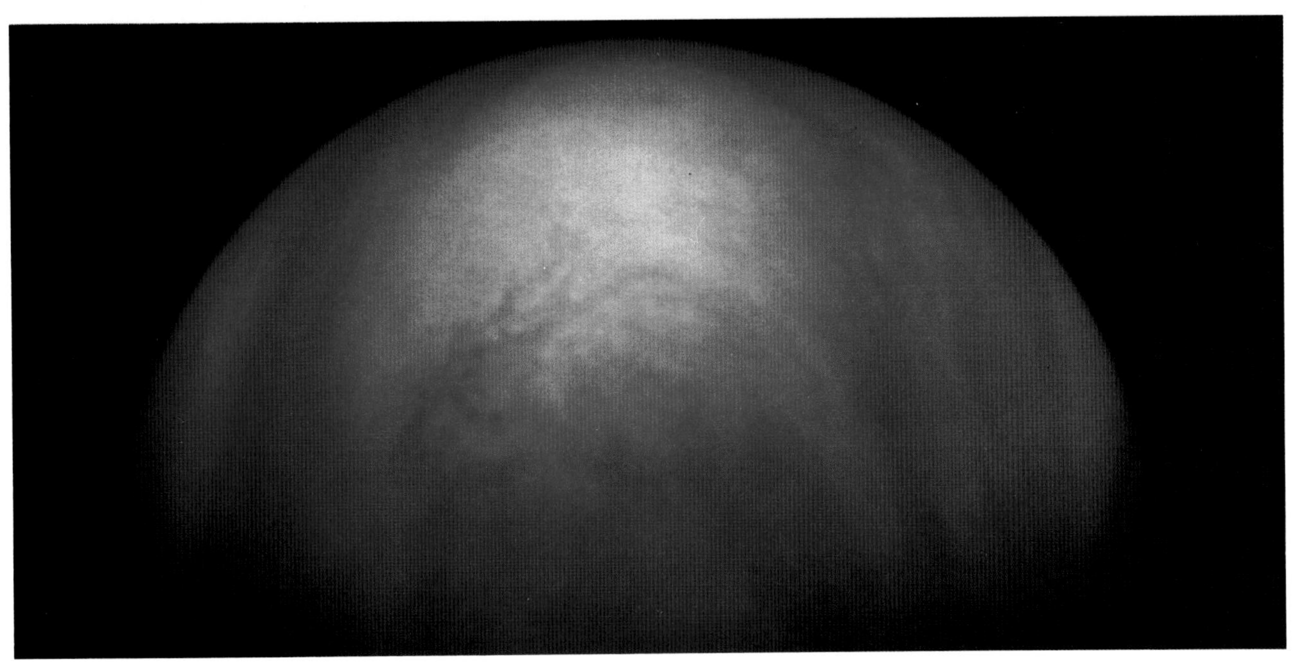

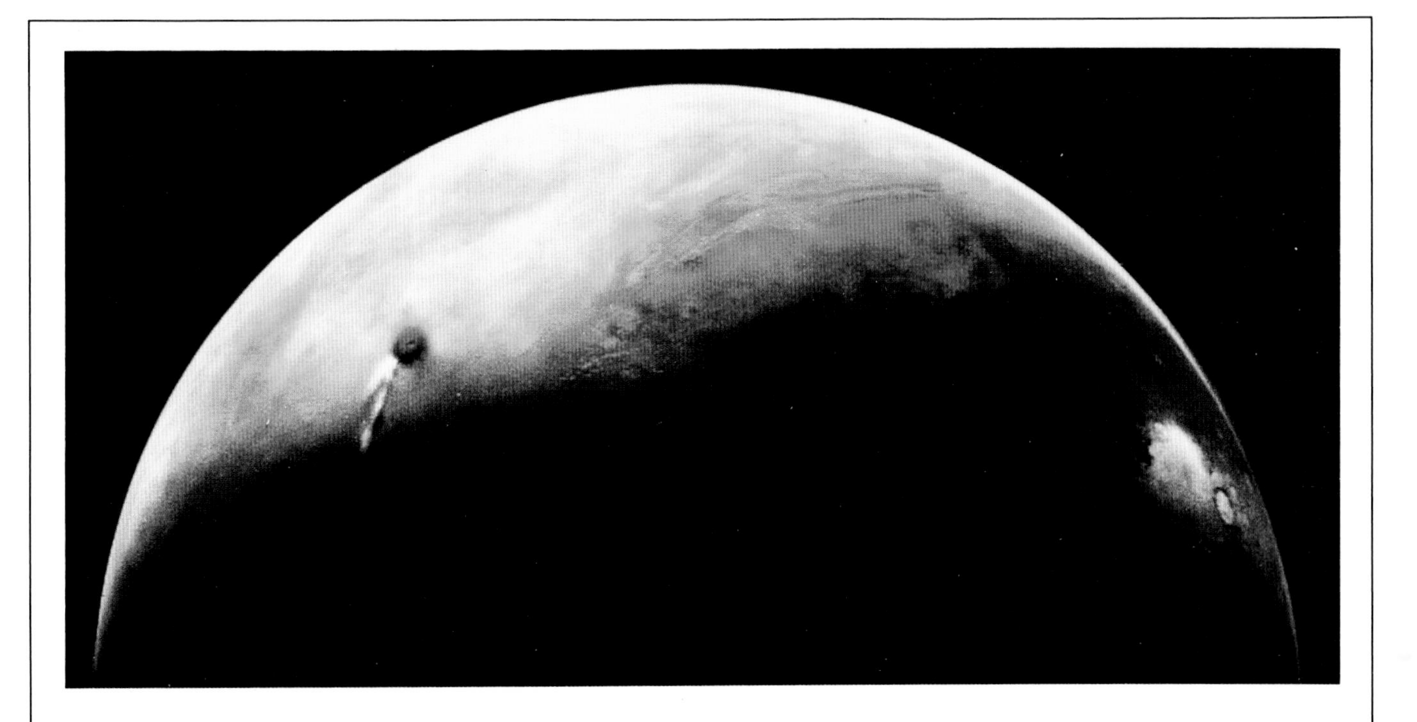

Page 68: Venus (computer-simulated).

Page 69: Top and bottom; Mars (computer-enhanced).

Page 70: Jupiter and its four moons (computer-collaged).

Page 71: Top; Saturn, its rings, and its major satellites (computer-collaged).
Bottom; Saturn's ring system (color-enhanced).

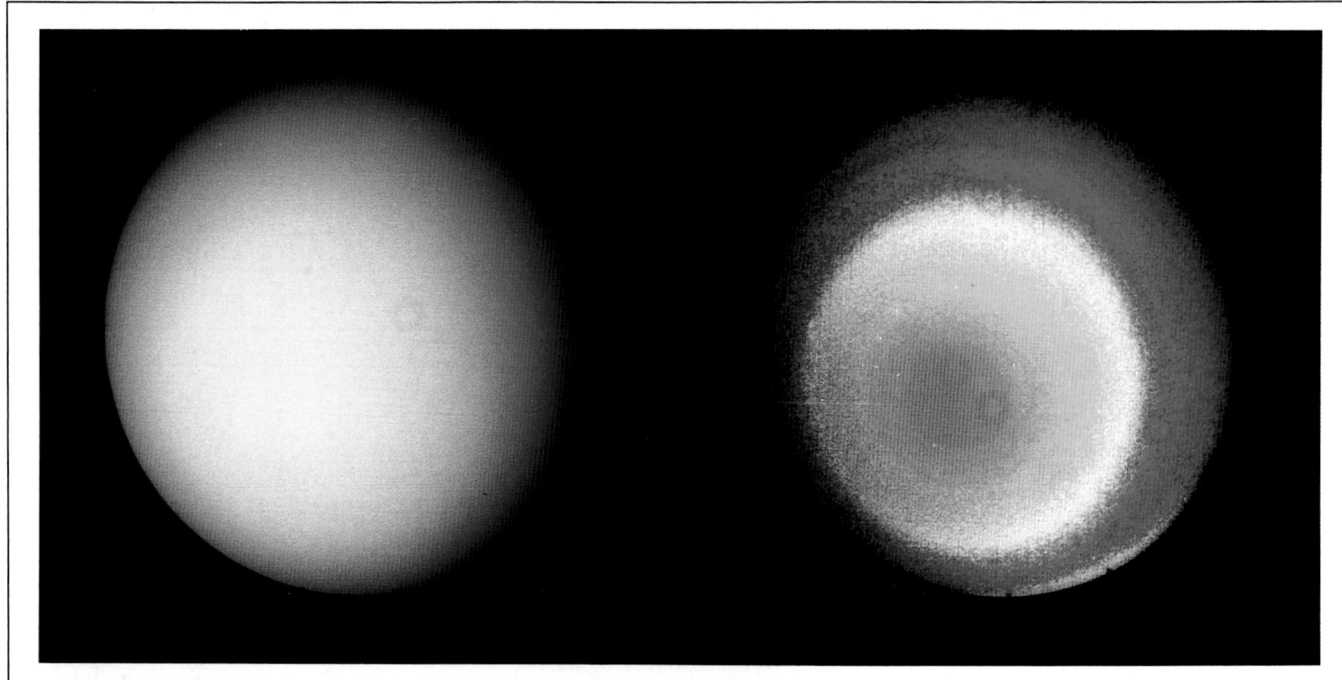

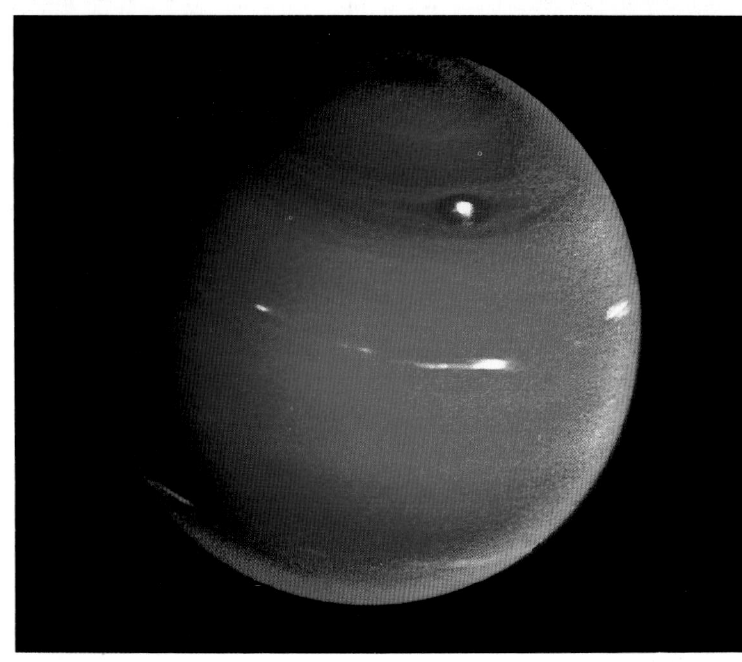

Page 72: Top; Uranus in true (left) and false (right) color. Bottom: Neptune (false color).

SECTION 4

THE INNER PLANETS

8 MERCURY AND VENUS: THE INNERMOST PLANETS

LEVEL: Basic

EQUIPMENT: Star maps, pencil, sighting stick

PROJECT: Determine by observation the angle of the ecliptic with respect to the horizon, and use this knowledge to predict what time of year is best for viewing Mercury and Venus

In addition to emitting steadier light, planets can be distinguished from stars by their colors. Venus, for example, is a pure, usually very bright white. In fact it can become so bright that many people cannot believe it is a planet. This brightness results from constant cloud cover that reflects more than 65 percent of the sunlight that hits Venus back into space.

Because Mercury is so close to the Sun, it is quite elusive and can only be seen well at certain times and under very good conditions over a low horizon. Venus is, in general, more visible. Viewers in the tropics have a better chance of seeing Mercury, because of their position relative to the ecliptic. Mercury can be more difficult to spot from high northern or southern latitudes, but with a little patience you should be able to find it in your evening or morning sky. The key is to figure out when it is at the highest angle with respect to your horizon in the early evening or just before sunrise. At these times, if Venus and/or Mercury is above the appropriate horizon, you should be able to see either or both planets clearly.

Take a look at your star maps and note that as the months pass, the angle between the ecliptic and the horizon increases and decreases. For mid-northern latitudes in the early evening, this angle is especially low from June through November and is higher from January through June. The reverse is true for the morning sky. The highest and lowest times occur in the middle of these ranges.

According to myth, the Greek god Mercury had wings on his feet and was known for his speed. It is no coincidence that the planet which makes the fastest orbit around the Sun is named Mercury.

Mercury can only be seen easily at the most favorable of its greatest elongations—the times when, as seen from Earth, it is farthest from the Sun. The months around Mercury's greatest elongations for the evening sky through the year 2000 are provided in Table 9, which indicates the constellation in which Mercury will appear in the month ahead when it is visible. See if you can figure out when the elongations will happen in the morning sky. More important, determine the best times to observe Mercury by comparing the months of its greatest elongation with the times when the ecliptic is highest.

Figure 35

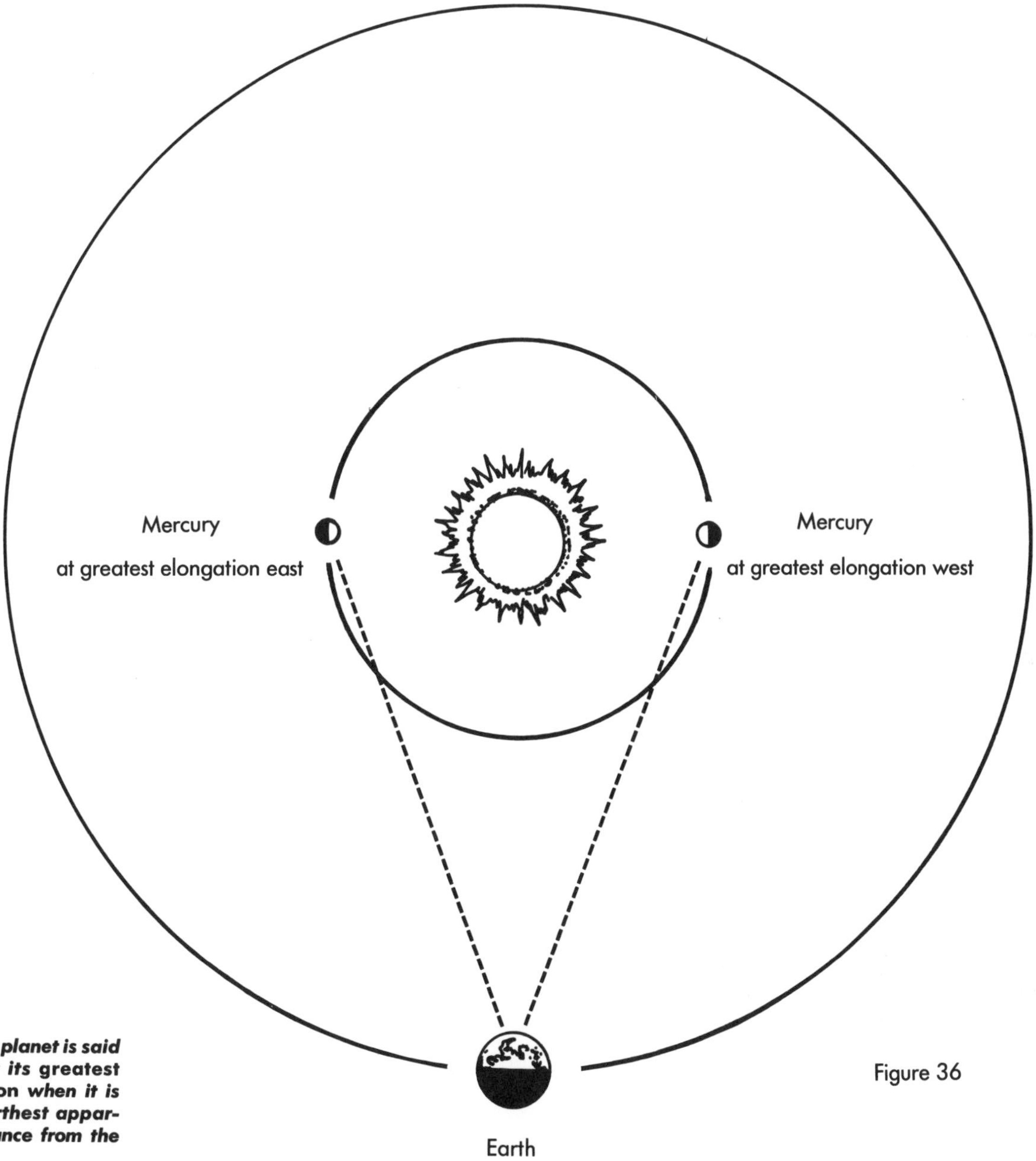

Mercury

at greatest elongation east

Mercury

at greatest elongation west

Earth

An inner planet is said to be at its greatest elongation when it is at its farthest apparent distance from the Sun.

Figure 36

TABLE 9
Position of Mercury along the Ecliptic

Year	Month(s) Mercury is near greatest eastern elongation	Constellation(s)
1993	February	Aquarius
	June	Gemini
	October	Libra
1994	January–February	Aquarius
	May–June	Taurus, Gemini
	September	Virgo
1995	January	Capricornus
	May	Taurus
	September	Virgo
1996	January	Sagittarius, Capricornus
	April	Aries
	August	Leo, Virgo
	December	Sagittarius
1997	April	Aries
	July–August	Leo
	November–December	Scorpius
1998	March	Pisces
	July	Cancer, Leo
	November	Scorpius
1999	February–March	Pisces
	June–July	Gemini, Cancer
	October	Libra
2000	February	Aquarius
	June	Gemini
	September–October	Virgo, Libra

Mercury makes very low and brief appearances alternately in the morning and evening skies.

Try to find Mercury in the sky at what you predict will be the best times for viewing. If you have a telescope, try Project 9 once you have found this difficult planet. If you can't find it without the aid of a telescope, realize that most people never see Mercury in their entire lives.

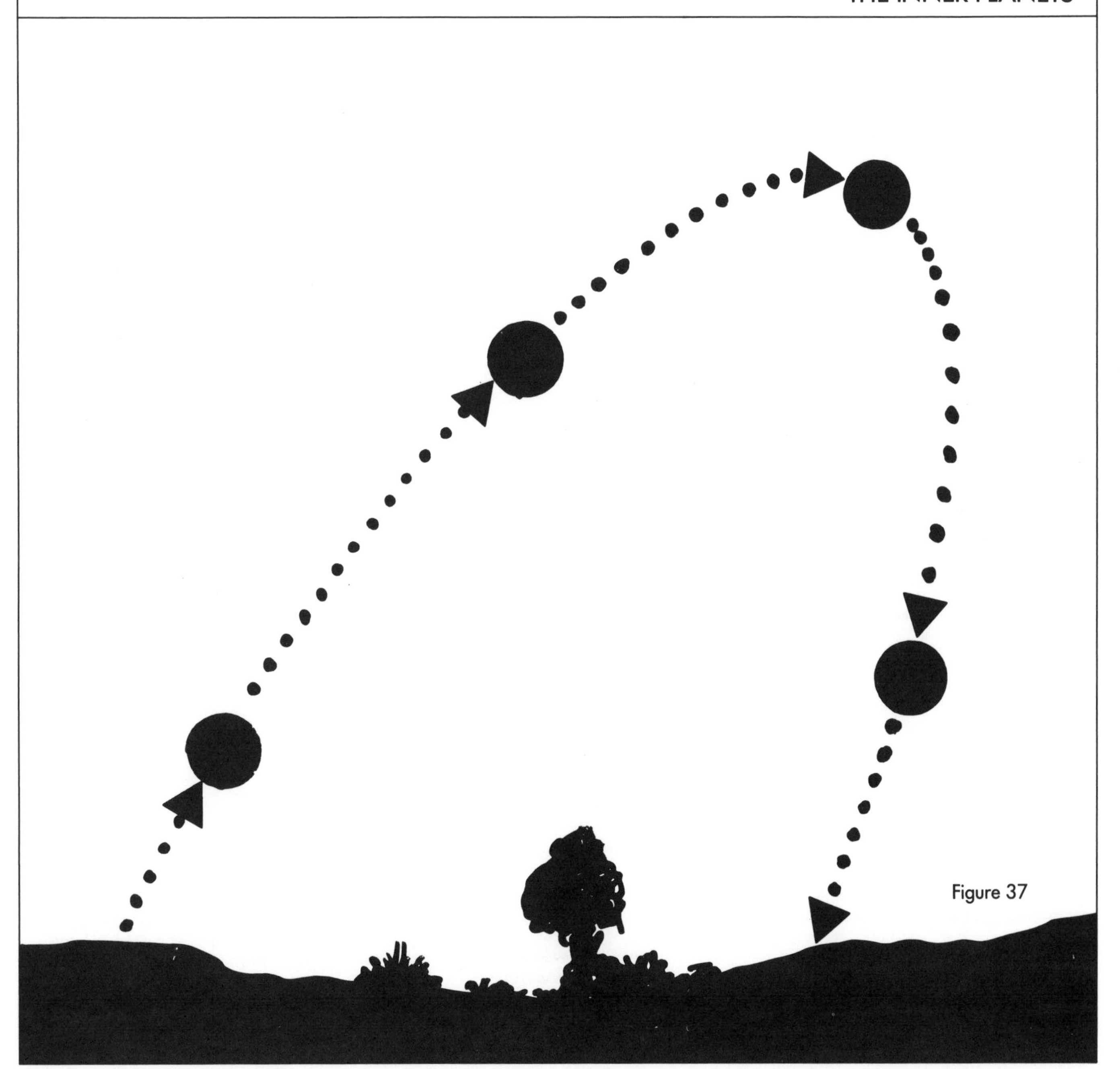

Figure 37

9 MERCURIAL MERCURY

LEVEL: Intermediate

EQUIPMENT: Telescope on tripod or mount, information from Project 8 or knowledge of where Mercury can be seen

PROJECT: Over a period of time around a favorable elongation of Mercury, find the planet and observe its phases; draw it if you wish

Mercury is an elusive planet. Like the Moon, it displays different phases, because it passes closer to the Sun than does Earth. However, you cannot view all of Mercury's phases, because it *is* so close to the Sun. Mercury is easiest to see when it is a thin crescent, but unfortunately it is never very high in the sky at night. Only at the

Because Mercury is closer to the Sun than Earth is, it exhibits phases similar to the Moon's.

Figure 39A Gibbous

most favorable elongations can you see it easily. These favorable elongations are times (as you will discover by doing Project 8) when Mercury is not simply farthest from the Sun (which happens at any greatest elongation) but is also at a point high enough above the horizon to be easily observed. Even then, you may be able to see it often only if you have a very flat horizon and make your observation within thirty minutes after sunset or before sunrise. Mercury is close enough, however, that if you can spot it, you will be able to see a lot with a low-power telescope or high-power spotting scope. You may even be able to make out some detail. High power is really not the best way to view Mercury. You should be able to see it with less than 100-power magnification. Because it is low

With a telescope and excellent viewing conditions, you can see slight details on Mercury.

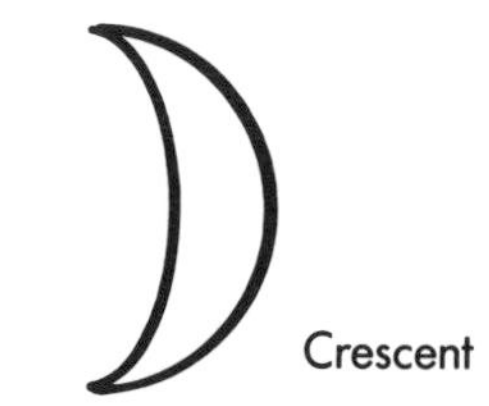

Figure 39B Crescent

in the sky, increasing power magnifies the effects of atmospheric turbulence. (When you look at a celestial object that's close to the horizon, the angle means you're looking through a lot more of Earth's atmosphere than when you look at an object that's higher in the sky.) Also, when Mercury is darkest, it is lower in the sky. All these factors together present a good challenge to your observing skills.

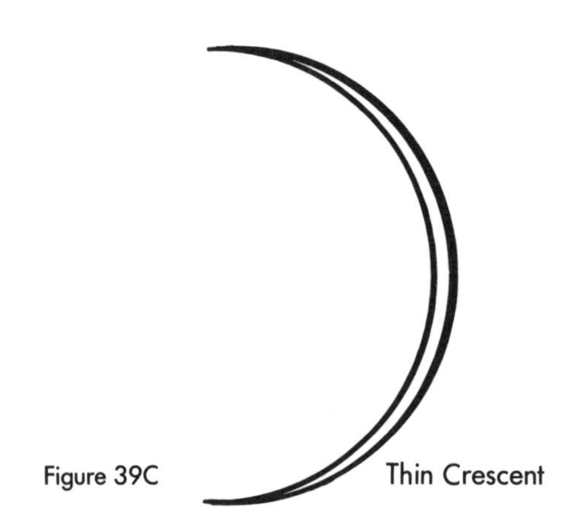

Figure 39C Thin Crescent

Although Mercury's coloration is similar to the Moon's, its color changes and may appear as yellowish, a pinkish-rose, or some other color. You may be able to see faint dark markings on the planet and possibly even some bright ones if conditions are good. You should have at least a 4-inch diameter telescope if you expect to see markings. A larger aperture is preferable, but

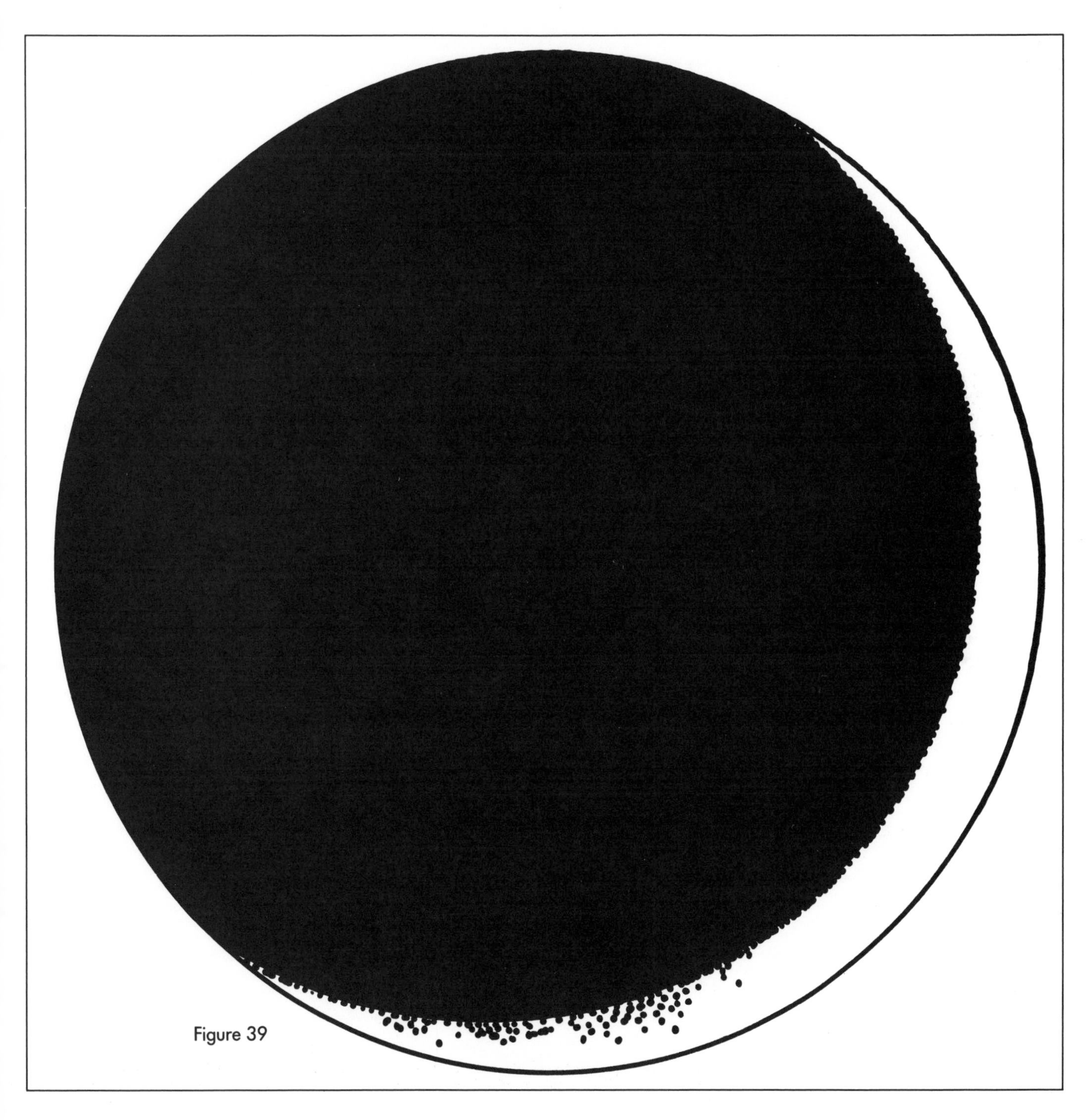
Figure 39

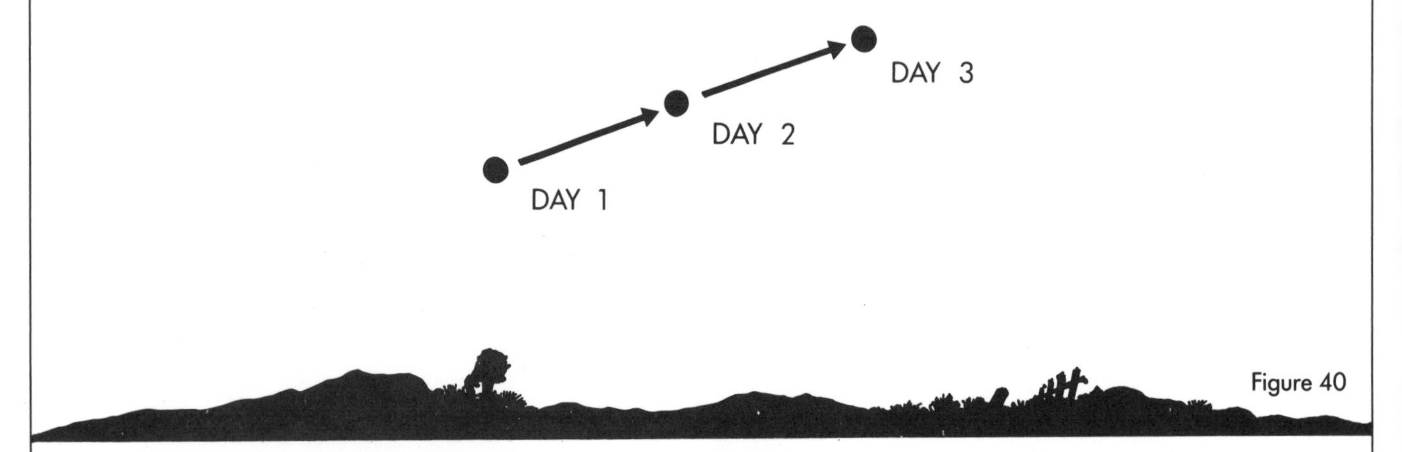

Figure 40

remember to stay with a low magnifying power. Record how Mercury looks to you and the exact times when you first see it and when you can last see it. A fun activity to try is seeing how long (and how low) in the sky you can view Mercury with the unaided eye, and then to compare how much longer you can track it with your telescope. It will eventually disappear behind something or be lost because of atmospheric extinction.

If you can get to the right location, you will be able to view two *transits* of Mercury during the 1990s. These are, in effect, partial eclipses of the Sun, caused by Mercury. But because the planet is 150 times smaller than the Sun, its transits go unnoticed unless you are prepared to look at the Sun properly, at the right time and place. On November 5, 1993, and November 15, 1999, Mercury will transit the Sun. Both events will be visible from Australia and New Zealand; the former will also be visible from China, Japan, Indonesia and India, and the latter from Hawaii and the west coast of North and South America, as well as large parts of the Pacific Ocean.

The same rules of observation that apply to solar and lunar eclipses can be applied to observing Mercury's transits. The points when Mercury first appears against the solar disk, when it is fully visible, when it starts to disappear, and when it has disappeared completely are of great interest to modern astronomers. Astronomers still have some refining to do in calculating the orbit of Mercury, and every additional observation helps. For your observation to have any importance to astronomers, accurate timings and your exact location at the time are essential.

Because of Mercury's rapid orbit around the sun, it changes position in the sky rapidly.

When Mercury appears in the evening sky, it is only visible for an hour, at best, after sunset.

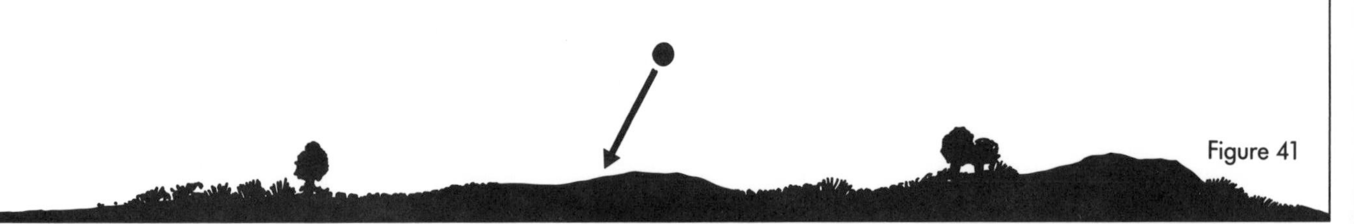
Figure 41

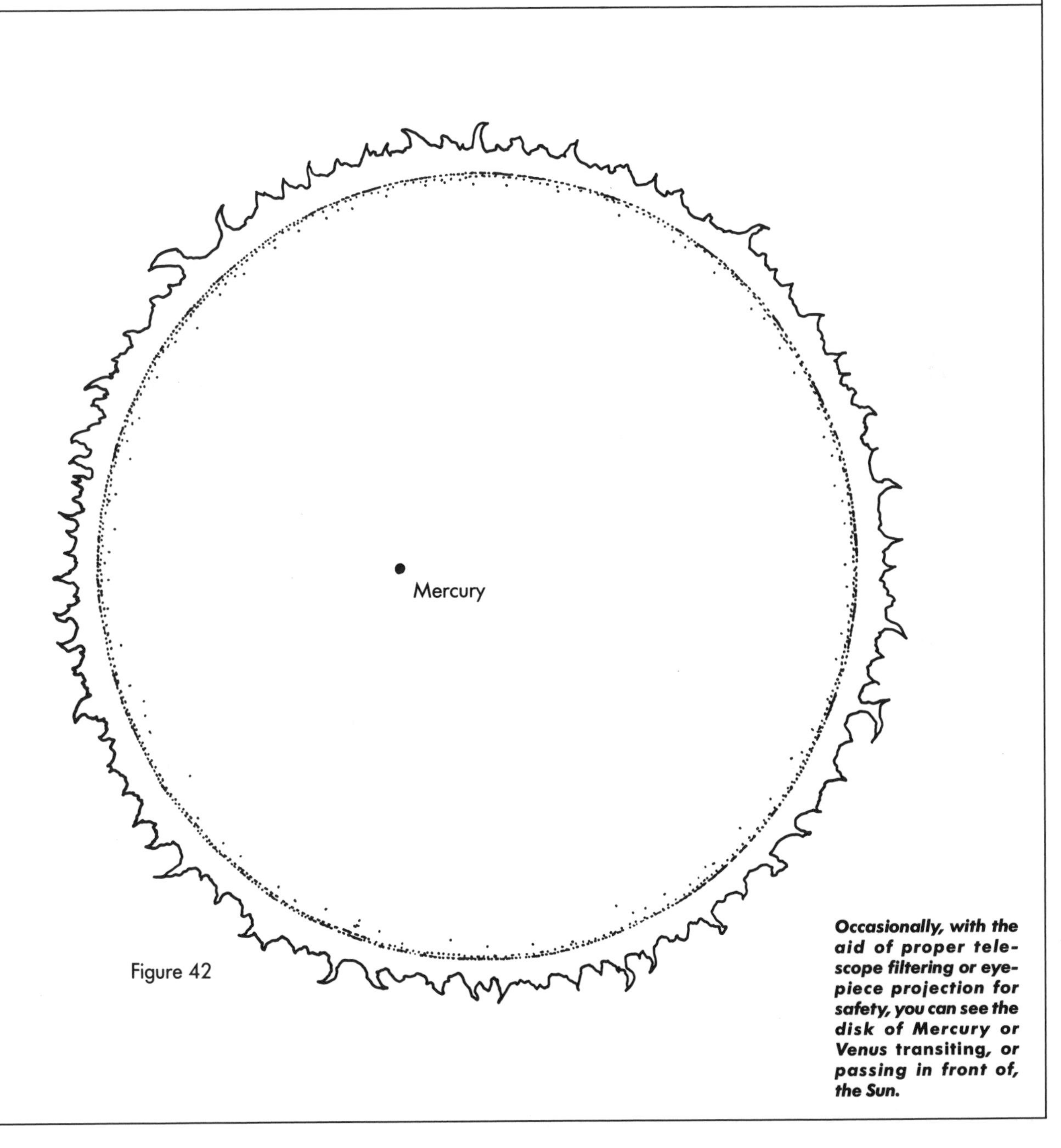

Figure 42

Occasionally, with the aid of proper telescope filtering or eyepiece projection for safety, you can see the disk of Mercury or Venus transiting, or passing in front of, the Sun.

10 VENUS, THE EVENING AND MORNING "STAR"

LEVEL: Basic

EQUIPMENT: Planet location chart for Venus, telescope or spotting scope on mount or tripod (optional)

PROJECT: Soon after sunset, or even before, locate the planet Venus in the evening twilight and/or see how long after morning twilight Venus remains visible; observe and record Venus's phases if optical aid is available

Because Venus can get so amazingly bright, it is known as "the number one UFO on record." At times it shines brilliantly, either in the evening long before any other celestial objects can be seen, or in a fairly bright blue sky in the morning. In fact, once you have found it, Venus can often be seen easily during the day. You simply have to know exactly where to look.

Figure 43A Full

Like Mercury, Venus goes through phases that can be easily observed using a small telescope. If you have very keen eyesight, you may be able—just barely—to see the phase of Venus when it is almost at its closest to Earth. Powerful binoculars are necessary to reveal the crescent phases, and you'll need more power to see the quarter, gibbous, and full phases. Interestingly, since Venus's distance from the Earth varies so much, it reaches its point of greatest brilliance in the crescent phase and can most easily be seen then. Disappointingly, although Venus is very bright, not much detail is visible, even with a telescope. Record the phase you see and any details, which you may wish to draw. It is extremely difficult to see such details without exactly the right conditions and a fairly large telescope, so be sure that what you are seeing

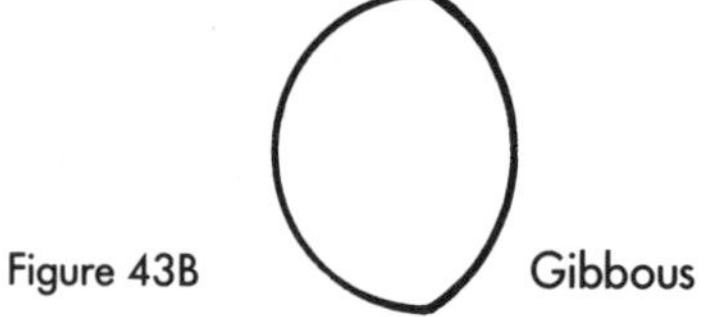
Figure 43B Gibbous

Like Mercury and the Moon, Venus exhibits full, gibbous, quarter, and crescent phases and appears larger and smaller depending on its distance from Earth.

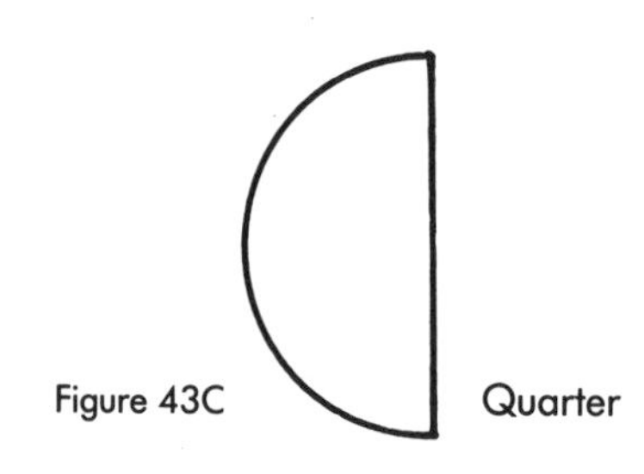
Figure 43C Quarter

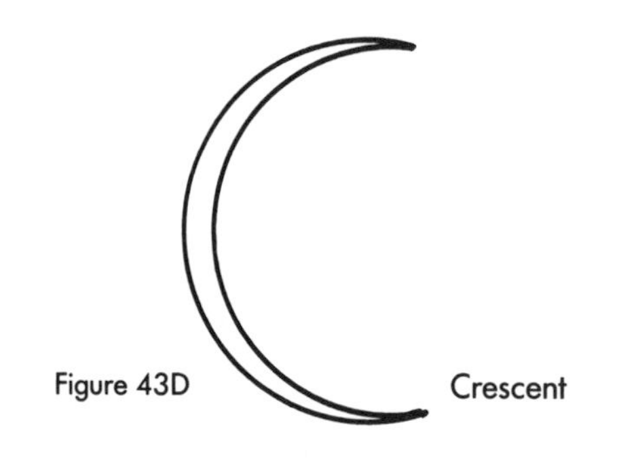
Figure 43D Crescent

Figure 44

is actually on the planet and not a problem with your instrument.

The knowledge you gain in Project 8 will help you determine the best times to observe Venus for the longest period on a given night (when the angle between the ecliptic and horizon is greatest). Table 10 will help you determine when the planet is in the morning or evening sky, using the same methods that were used for Mercury.

Venus, the Roman goddess of beauty, is the namesake of the brightest planet.

An accurate time source is vital to making useful astronomical observations.

TABLE 10
Position of Venus in the morning and evening sky

Year	Month(s) in evening sky	Month(s) in morning sky
1993	January–March	April–October
1994	March–October	December
1995	November–December	January–May
1996	January–May	July–December
1997	June–December	Not visible
1998	Not visible	February–August
1999	January–August	September–December
2000	August–December	January–March

At times not shown on Table 10, Venus is, for the most part, too near the solar glare to be visible. See if you can spot it early or late in the day as the dates in the table approach or pass. You will notice that, as compared with Mercury, Venus stays in the sky for much longer stretches. It is nearly twice as far as Mercury from the Sun and takes more than two and a half times longer to orbit the Sun.

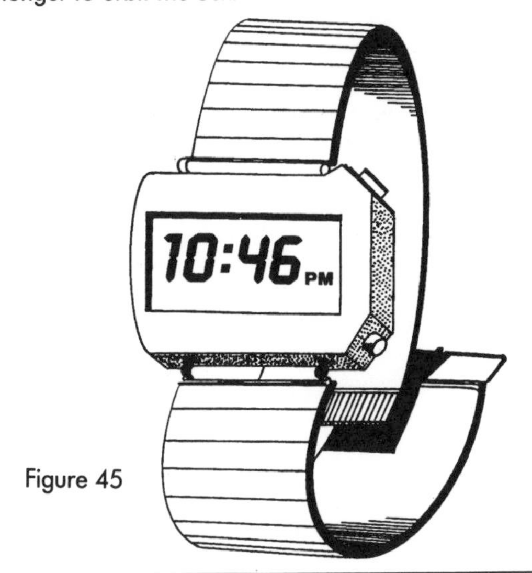

Figure 45

11 VENUS, THE BRIGHTEST PLANET

LEVEL: Intermediate

EQUIPMENT: *Astronomical Almanac*, star chart with brightness indications of stars

PROJECT: Using information on the brightnesses of nearby stars, determine how bright each planet is at the time you observe it

One of the confusing things about viewing planets is that there is no completely firm rule as to which is brightest in the sky. Venus is generally the brightest—if seen at all—and Saturn and Mercury are usually fainter than the other planets that are visible to the unaided eye. Still, it can be difficult to determine exactly which planet you are seeing, particularly if you are mildly color-blind or your eyes are not quite sensitive enough to discern the rather subtle colors of the night sky. Here we will identify planets by collecting data on the order of brightness, and along the way we'll discover some patterns useful not only in identifying a particular planet but also in figuring out the best time to look for it.

To begin this project, you need to become familiar with the sources of information on the brightness of planets. The *Observer's Handbook* provides listings for several dates throughout the year. The *Astronomical Almanac* is more complete, however, with lists for each planet about every five days (Table 11). These listings are by *magnitude*, which is the brightness standard used by astronomers and on star charts. You will use this information later to check your estimates.

The magnitude scale is a bit difficult to get used to. First, it uses both negative and positive numbers. Magnitude 0 refers to many of the brighter stars, but negative magnitudes are even brighter than these, and larger positive magnitudes denote fainter stars. The stars Aldebaran (in the constellation of Taurus) and Altair (in Aquila) have magnitudes of +1, while Sirius (in Canis Major) is the brightest star at about −1.5. The full moon has a magnitude of −12.5, and the Sun −26.5. Under the best conditions with good eyes, you should be able to see to magnitudes to +6, which makes the planet Uranus (at magnitude +5.6) just barely visible. With average binoculars, you can see celestial bodies down to magnitude +10, which allows you to see Neptune (+7.9), and with an 8-inch telescope, you should be able to see Pluto (at nearly magnitude +14).

The other confusing aspect of magnitude is that the scale is not linear, but logarithmic. To aid in your estimates of the brightness of the planets, remember that if one object is twice as bright as another, its magnitude is 0.75 less. If it is four times as bright, its magnitude decreases by 1.5. If it is ten times brighter, the magnitude decreases by 2.5, and so on. To start observing, you will need a star chart (such as those in *Sky and Telescope* magazine) that shows a magnitude scale with good resolution. An even better resource is the *Sky Atlas 2000*, by Wil Tirion. This book is extremely useful in many ways if you are planning to observe the sky in serious detail. The more detailed a chart you can find, the better, but unless you have access to a good university library, highly detailed charts will be difficult to find (and they're also very expensive).

Once you have your star chart, use it to find out where a particular planet should be. With Venus, Jupiter, and sometimes Mars, it is difficult to estimate brightness, because the former two are often, and the latter is occasionally, brighter than anything else in the sky besides the Sun and Moon. There are, therefore, no good "comparison stars" of known brightness with which to compare these planets. When Sirius is in the sky, you can use it as a standard to estimate for other stars, but at a magnitude of about −1.5, it is a bit dim to compare with the

−4.6 magnitude that Venus can sometimes reach. But, using the information above, do try to estimate the magnitude of Venus. Is Venus twice as bright as your comparison star? Three times as bright? Of course it is best to find a star that most closely approximates the brightness of the planet you are observing. This is a little easier with Mars, Saturn, and Mercury.

This method of learning about the stars and making comparisons can be useful in other ways, too. Try applying it to observing comets and meteors (see Project 17). A bright comet, for example, may have a very bright center.

Venus is almost always the brightest planet. Mars and Jupiter sometimes tie for second, with Saturn and especially Mercury somewhat fainter.

Make magnitude estimates of this concentration of light. Even with fainter comets, especially if they are visible with the unaided eye, you can compare them with stars and make rough "total magnitude" estimates. A comet looks like a fuzzy patch, but it appears roughly as bright as a star. See what you can see.

As you learn more about the planets, you will come to recognize at a glance exactly which one you are seeing, either by its color, position, change in motion or brightness. It won't be long before you begin picking them out as easily as constellations.

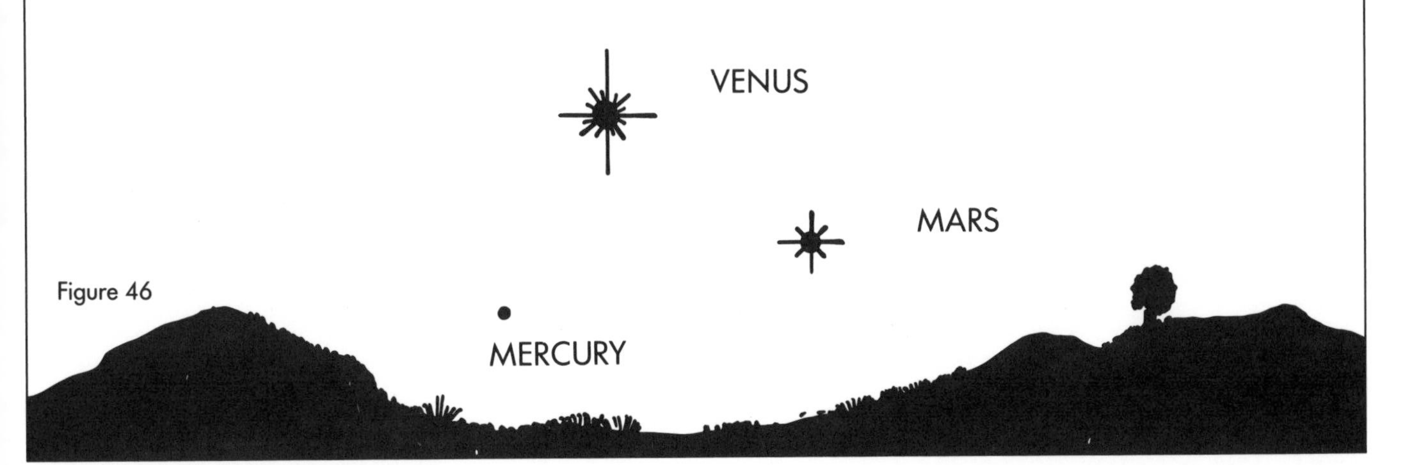

Figure 46

11 VENUS, THE BRIGHTEST PLANET

TABLE 11

PHENOMENA, 1989

ELONGATIONS AND MAGNITUDES OF PLANETS AT 0^h UT

	Mercury		Venus	
Date	Elong.	Mag.	Elong.	Mag.
Jan. −1	E. 16°	−0.7	W. 23°	−3.9
4	18	0.7	22	3.9
9	19	−0.6	21	3.9
14	18	0.0	20	3.9
19	12	+1.6	19	3.9
24	E. 4	+4.4	W. 17	−3.9
29	W. 10	2.9	16	3.9
Feb. 3	18	1.2	15	3.9
8	23	0.5	14	3.9
13	26	0.2	13	3.9
18	W. 26	+0.1	W. 11	−3.9
23	26	0.0	10	3.9
28	25	0.0	9	3.9
Mar. 5	23	−0.1	8	3.9
10	21	0.2	7	3.9
15	W. 18	−0.4	W. 5	−3.9
20	14	0.6	4	3.9
25	10	0.9	3	3.9
30	6	1.4	2	3.9

	Mercury		Venus	
Date	Elong.	Mag.	Elong.	Mag.
Apr. 4	W. 1°	2.0	W. 1°	3.9
9	E. 5	−1.8	E. 2	−3.9
14	10	1.4	3	3.9
19	15	1.0	4	3.9
24	19	−0.5	5	3.9
29	21	+0.1	6	3.9
May 4	E. 20	+0.7	E. 8	−3.9
9	18	1.6	9	3.9
14	14	2.8	10	3.9
19	E. 7	4.3	12	3.9
24	W. 2	5.8	13	3.9
29	W. 8	+4.2	E. 14	−3.9
June 3	15	2.8	16	3.9
8	19	1.8	17	3.9
13	22	1.1	18	3.9
18	23	0.6	20	3.9
23	W. 22	+0.1	E. 21	−3.9
28	20	−0.3	22	3.9

The Astronomical Almanac, ***in which this chart appears, and the*** **Observer's Handbook,** ***are annual publications that provide important information on planet positions and brightness.***

Date		Mercury Elong.	Mercury Mag.	Venus Elong.	Venus Mag.
July	**3**	W. 17°	−0.8	E. 23°	−3.9
	8	12	1.2	25	3.9
	13	6	1.7	26	3.9
	18	W. 2	2.1	27	3.9
	23	E. 6	1.6	29	3.9
	28	E. 11	−1.0	E. 30	−3.9
Aug.	**2**	15	0.6	31	3.9
	7	19	0.4	32	3.9
	12	22	−0.2	33	4.0
	17	24	0.0	35	4.0
	22	E. 26	+0.1	E. 36	−4.0
	27	27	0.2	37	4.0
Sept.	**1**	27	0.3	38	4.0
	6	26	0.5	39	4.0
	11	23	0.9	40	4.0
	16	E. 17	+1.7	E. 41	−4.1
	21	E. 9	3.4	42	4.1
	26	W. 3	4.8	43	4.1

Date		Mercury Elong.	Mercury Mag.	Venus Elong.	Venus Mag.
Oct.	**1**	11°	2.1	44°	4.1
	6	17	+0.3	44	4.2
	11	W. 18	−0.5	E. 45	−4.2
	16	17	0.9	46	4.2
	21	14	1.0	46	4.3
	26	10	1.0	47	4.3
	31	7	1.1	47	4.3
Nov.	**5**	W. 4	−1.2	E. 47	−4.4
	10	0	1.3	47	4.4
	15	E. 3	1.1	47	4.5
	20	5	0.9	47	4.5
	25	8	0.7	46	4.6
	30	E. 11	−0.6	E. 45	−4.6
Dec.	**5**	13	0.6	44	4.6
	10	16	0.6	42	4.7
	15	18	0.6	40	4.7
	20	20	0.6	37	4.7
	25	E. 20	−0.4	E. 33	−4.6
	30	18	+0.2	28	4.6
	35	E. 11	+1.9	E. 23	−4.5

MINOR PLANETS

		Conjunction	Stationary	Opposition	Stationary
Ceres	...	Apr. 28	Nov. 3	Dec. 20	—
Pallas	...	Feb. 25	Aug. 18	Sept. 30	Nov. 23
Juno	...	Oct. 9	Jan. 4	Feb. 21	Apr. 5
Vesta	...	—	May 14	June 26	Aug. 7

12 MARS

LEVEL: Advanced

EQUIPMENT: Observation form, telescope, orange and blue pencils

PROJECT: Observe and draw features on the surface of Mars

The most striking thing about Mars is its reddish-orange color. It looks very much like (and should not be confused with) three stars: Aldebaran (in Taurus), Betelgeuse (in Orion), and Antares (in Scorpius). The name Antares actually means "rival of Mars." Mars can be seen near all three of these stars during the course of its journey around the Sun. Because it is much closer to Earth and orbits the Sun, however, it appears to move from night to night, and its light is steadier than that of the stars. Because Mars is only half the size of Earth, it actually appears quite small in binoculars and telescopes, even though it can get rather close to us at *opposition*, the time when a planet is on the other side of the sky from the Sun, directly opposite it. Opposition is when Mars can best be viewed. With high-power binoculars, you may barely be able to make out one of the white polar caps against the orangish disk. With a small telescope, unless there is a planetwide dust storm on Mars at the time, one or the other of the polar caps should be quite obvious. Over a period of a year or so, you will be able to see these caps change in size. As with the Earth's polar ice caps, the Martian polar caps grow and shrink with the seasons. The difference is that the seasons on Mars are longer than those on Earth, since it takes about 687 days for Mars to go around the Sun.

Observing Mars will take some getting used to. At first, you may only see the polar caps. After some observing practice, slightly bluish-green markings may become visible. With larger telescopes, with a steady atmosphere and the planet at opposition, you might be able to see a lot. Some early observers of the planet noted that the dark bluish-green markings seemed to change with the seasons, indicating that they might be vegetation. But space probes to Mars have confirmed that this is not the case. The planet is covered with craters, and although frozen water and water vapor have been found there, no signs of life have been detected. The dark markings seem to result from light-colored sands covering or uncovering darker underlying areas of the surface, and the apparent changes result from viewing conditions influenced by the atmosphere of Mars.

The more casual observer might find it interesting to plot the apparent path of Mars against the star background around the time of opposition. Shortly before opposition, as Earth begins to overtake Mars in Earth's smaller, faster orbit around the Sun, the red planet will seem to slow down and stop in the sky and then to "back up" for a while as we pass it. Then it appears to change direction again, returning to its east-to-west motion against the stars. This phenomenon is known as a *retrograde loop* and is shown in Figure 49.

If you are able to see features on Mars, you might try to draw them, using colored pencils to indicate the differently colored features. Because the rotation period of Mars (24.6 hours) is very much like that of Earth, you may be able to identify some of the same features from one night to the next.

The experienced observer can spot several major features on Mars (Figure 48). The more you practice observing, the more you will be able to pick out. Solis Lacus is a large circular looking feature, sometimes called "the eye of Mars" because it looks like the dark pupil of an eye when it appears near the center of the disk. See if you can find it.

The Tharsis region of Mars contains four huge volcanoes that appear as dark spots against the lighter surface. One might wonder

The apparent size of Mars (Figures 47 A and B) varies considerably with its distance from Earth. It can be as close as 50 million miles from Earth, or as far as 250 million miles away.

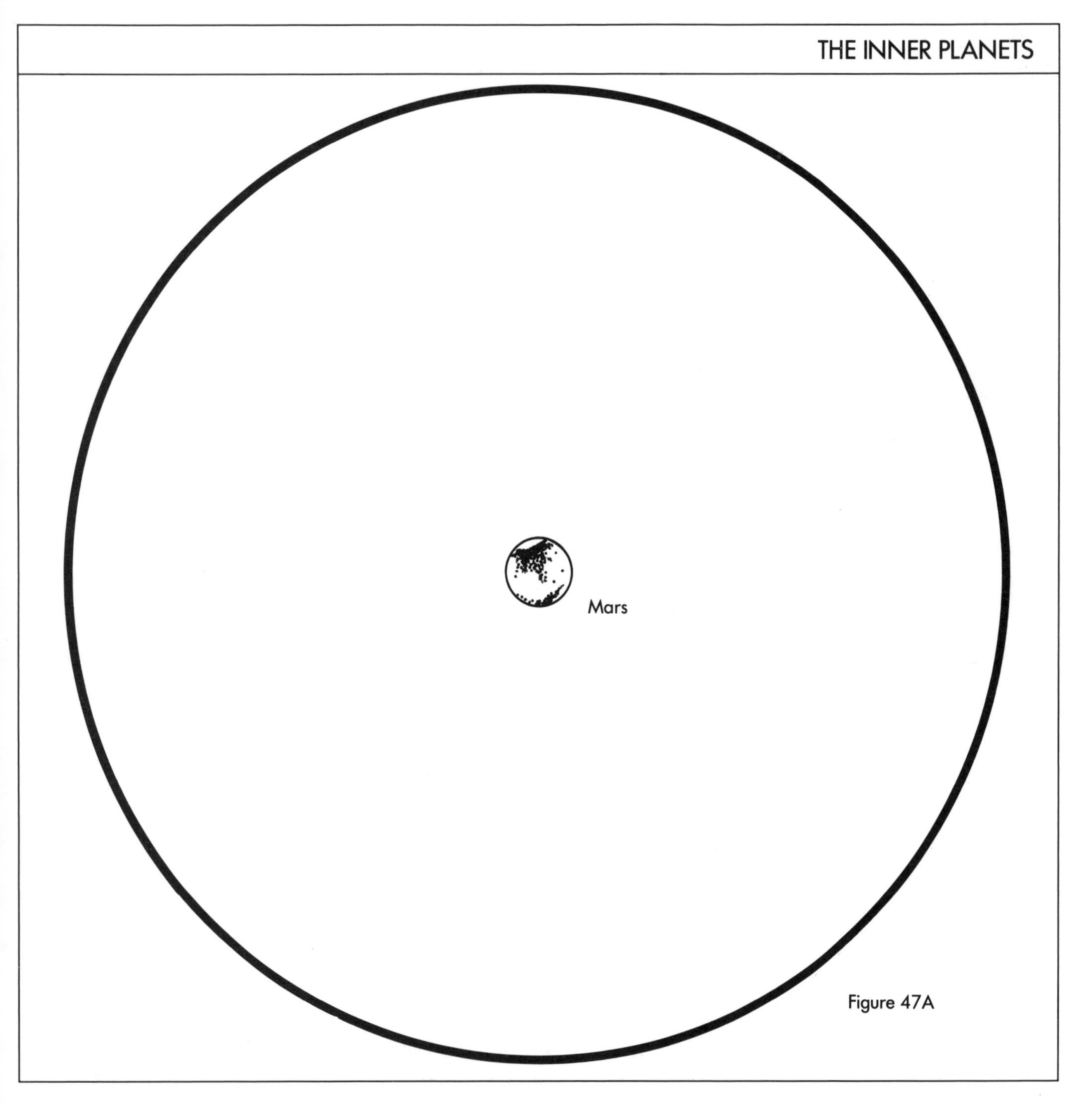

Figure 47A

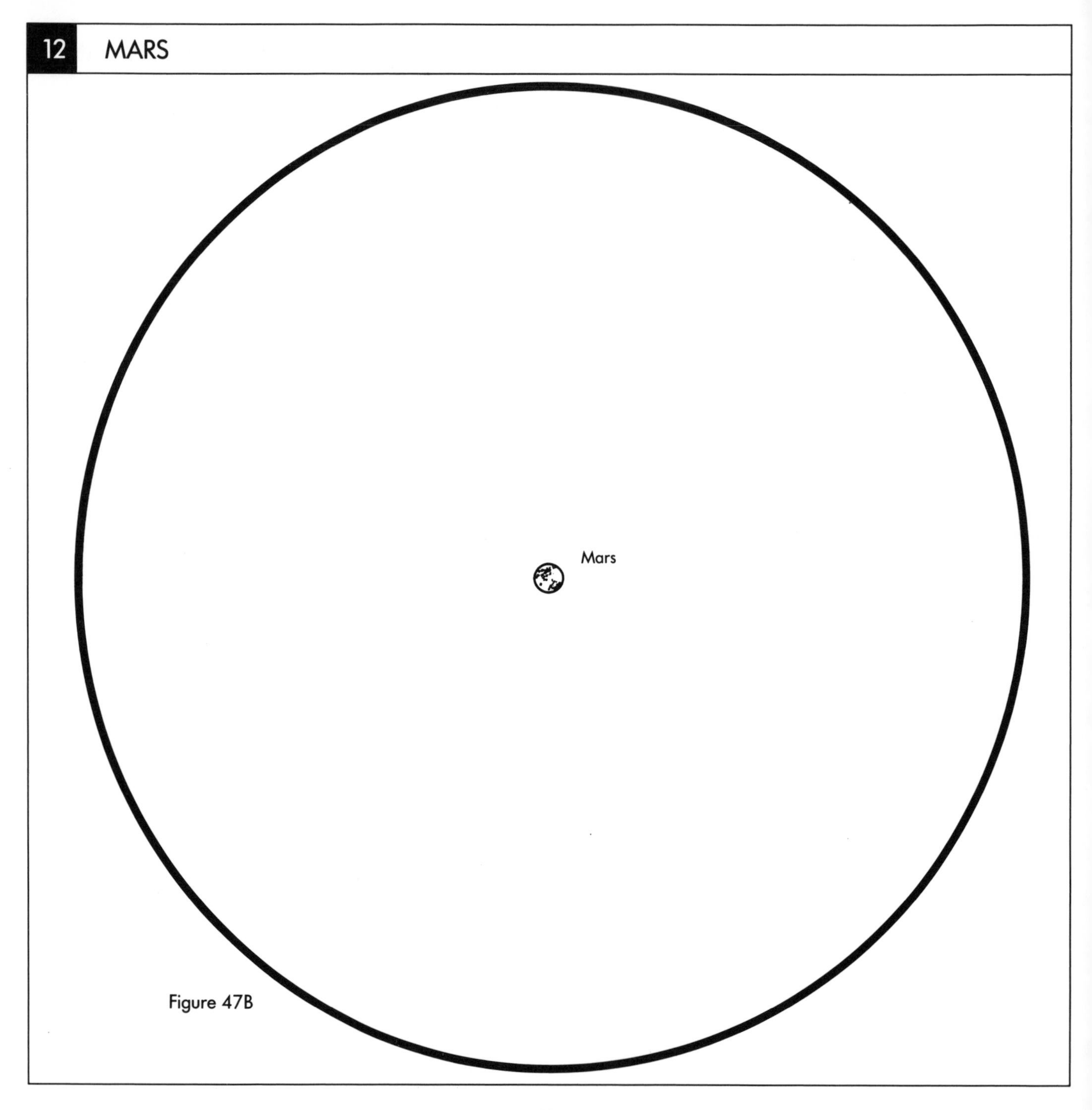

Figure 47B

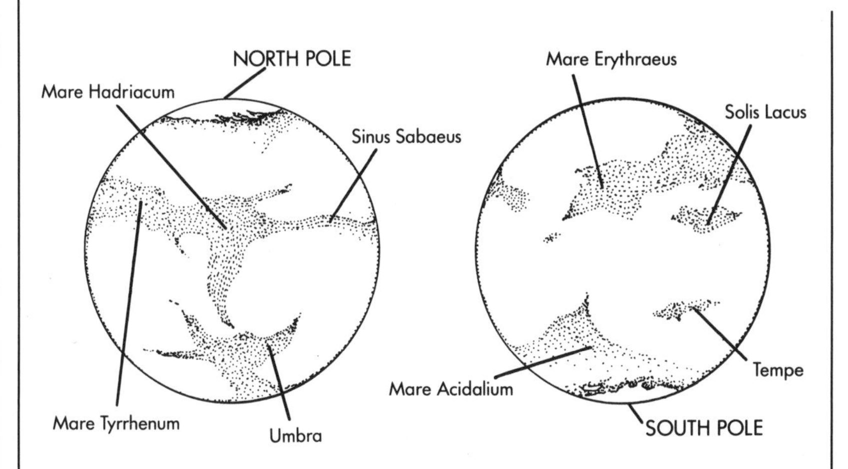

Figure 48

Seasonal heating and cooling cause the polar caps of Mars to vary in size. Although there is some ice on Mars, these polar caps are mainly carbon dioxide or "dry" ice. Many major features can be seen on Mars with medium-sized telescopes. Like the "seas" of the Moon, most have Latin names.

how volcanoes on Mars can be seen from the great distance of Earth. The reason is that they are quite large. The largest, Olympus Mons, is three times as high as Mount Everest, and if brought to Earth its base would cover the entire state of Utah.

Unfortunately, it is extremely difficult to see craters and many other features on Mars. Those huge volcanoes probably cast enormous shadows, and it's a shame we can't view them as the terminator passes by, which would make viewing Mars much more rewarding. (Remember that full moon is not the best time to see lunar features in detail. The only views we get of Mars are nearly "full," hence, many features are not observable, even in the largest telescopes. But there are many that *can* be seen if you look carefully.)

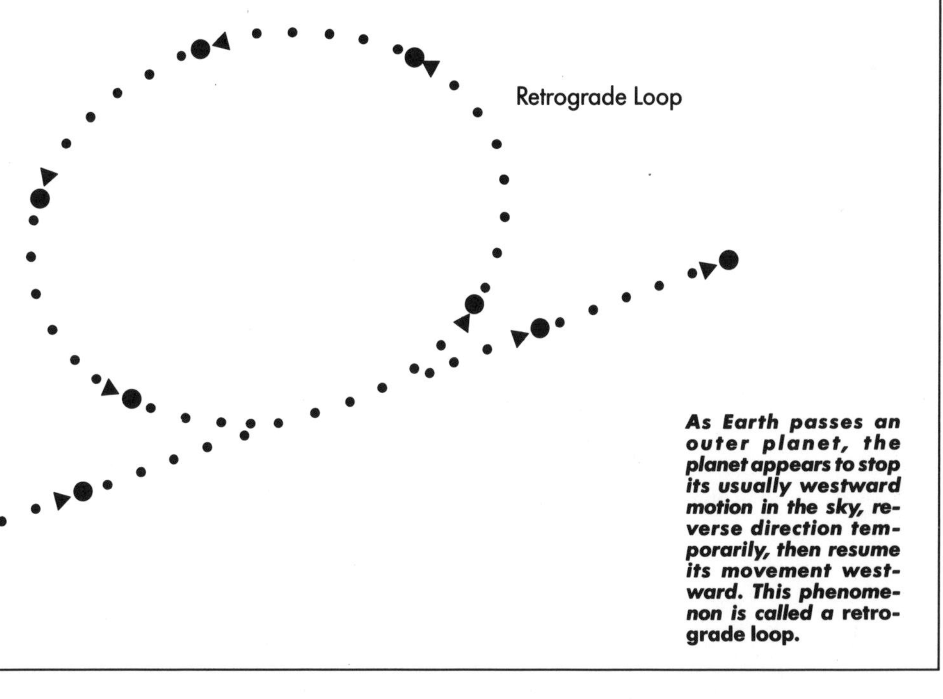

Figure 49

As Earth passes an outer planet, the planet appears to stop its usually westward motion in the sky, reverse direction temporarily, then resume its movement westward. This phenomenon is called a retrograde loop.

SECTION 5

THE OUTER PLANETS

13 JUPITER'S CLOUD BANDS

LEVEL: Advanced

EQUIPMENT: Telescope on mounting or tripod, preferably with a clock drive, pencils (colored if desired)

PROJECT: Observe and draw features on Jupiter

Jupiter is off-white in apparent color and can vary considerably in brightness depending on its distance from Earth. It never gets as bright as Venus, but its apparent size nearly rivals that of the second planet. As with Mars, Jupiter gets closest to us at opposition (when it's about 500 million miles away). Viewing conditions vary, but at eleven times the Earth's diameter (and with a volume that could accommodate thirteen hundred Earths), Jupiter is so large that even when it looks smallest, it is about the same size as Mars at its best.

Jupiter is actually a huge ball of gas (it and the other large outer planets are sometimes referred to as the "gas giants"). Even in small telescopes, two major dark cloud bands are always visible on Jupiter (Figure 51). They seem to remain basically the same color and distance from Jupiter's equator. See if you can spot them as you begin observing the planet. You may also be able to see some tiny, star-like objects around Jupiter; as many as four may be apparent at a time. These are the largest of at least sixteen moons circling Jupiter. That Jupiter has moons was first discovered by the Italian astronomer Galileo in 1609. Note their positions when you draw the planet. In Project 14, you'll focus more attention on the moons of Jupiter.

You may also notice that there is a slightly pinkish oval visible at times on the northern half of the planet. This is the Great Red Spot, three times the diameter of Earth, which is like an enormous, permanent hurricane in the planet's atmosphere. This giant storm results from the rapid rotation of Jupiter, the interaction of various chemicals in its atmosphere, and meteorological effects in the planet's upper atmosphere.

As you become more experienced in viewing Jupiter, you may be able to see more details, more banding, and even some white spots that look like much smaller versions of the Great Red Spot. You can try to draw these features in color, or use the following labels, which are accepted by the Association of Lunar and Planetary Observers for marking colors of planets:

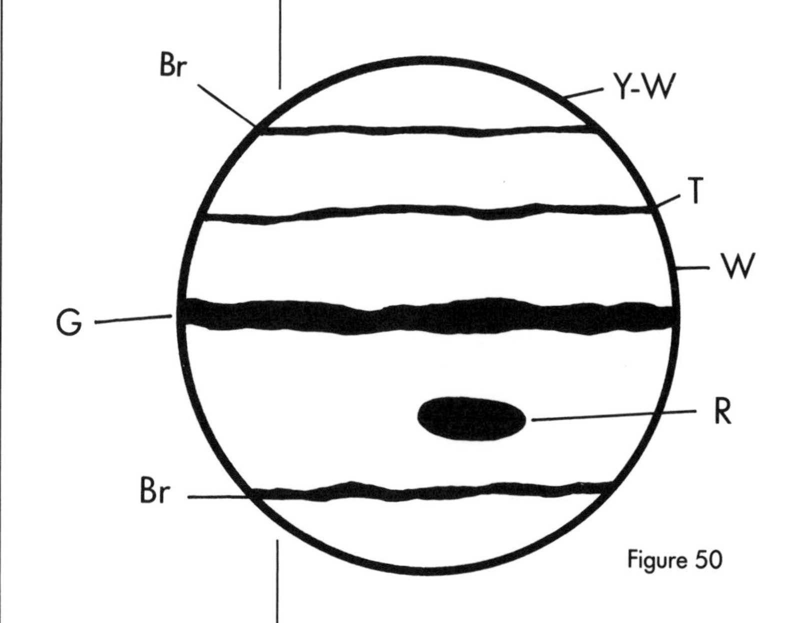

Figure 50

To the trained eye, several colors are apparent in the cloud bands and Great Red Spot of Jupiter.

TABLE 12
Planet Colors

W = white	O = orange
Y = yellow	Oc = ochre
Y-W = yellowish white	R = red
Bl = blue	T = tan
G = grey	Br = brown
Bl-g = bluish grey	R-Br = reddish brown

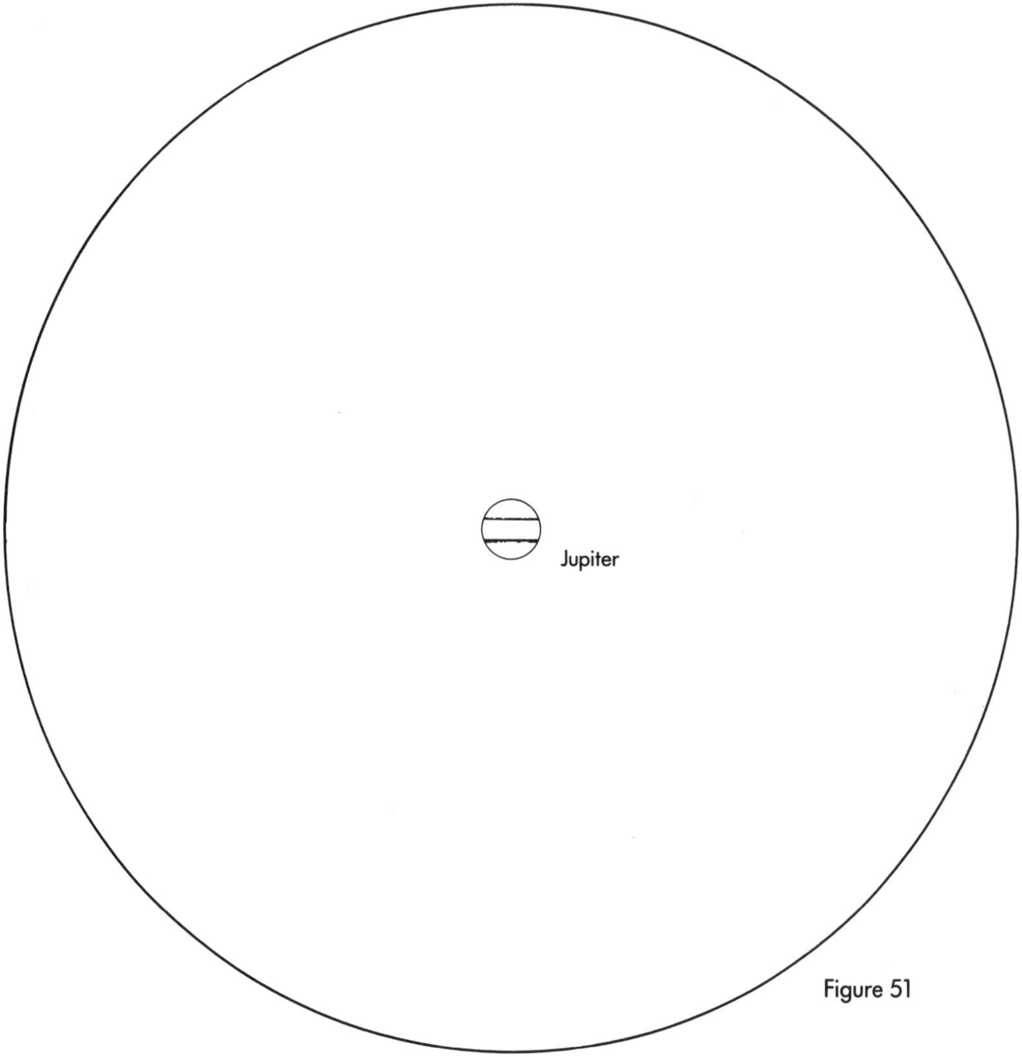

Figure 51

If you do see any localized features besides the Great Red Spot that are distinct enough to enable you to track their progress across the disk of Jupiter, make note of them and try to find them again later. Remember, though, that almost all features on this planet, except major cloud bands and the Great Red Spot, are transient. They come and go as atmospheric conditions change. Actually, even the Great Red Spot can change slightly in size, but at least it is always there.

13 JUPITER'S CLOUD BANDS

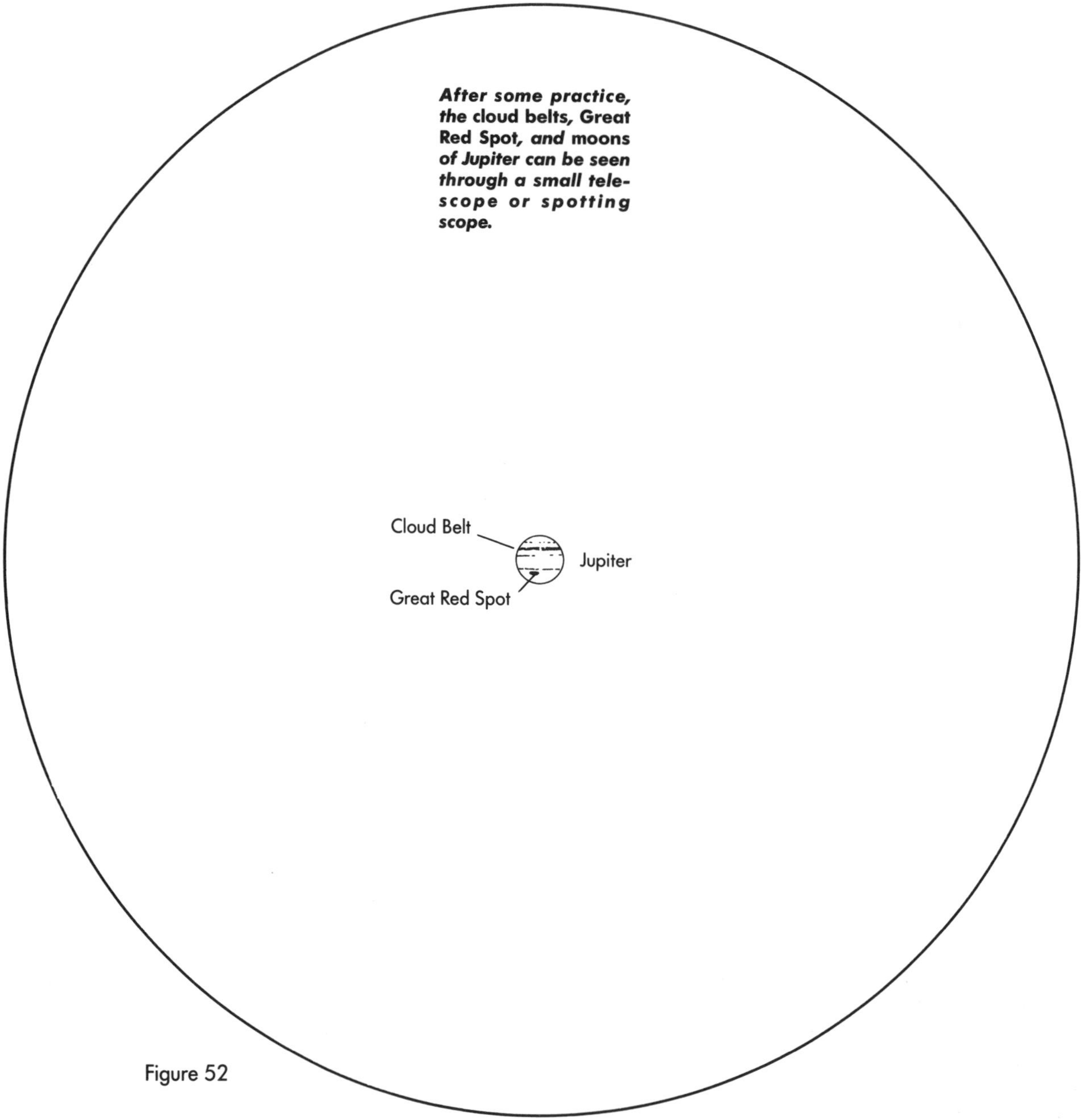

After some practice, the cloud belts, Great Red Spot, and moons of Jupiter can be seen through a small telescope or spotting scope.

Cloud Belt

Jupiter

Great Red Spot

Figure 52

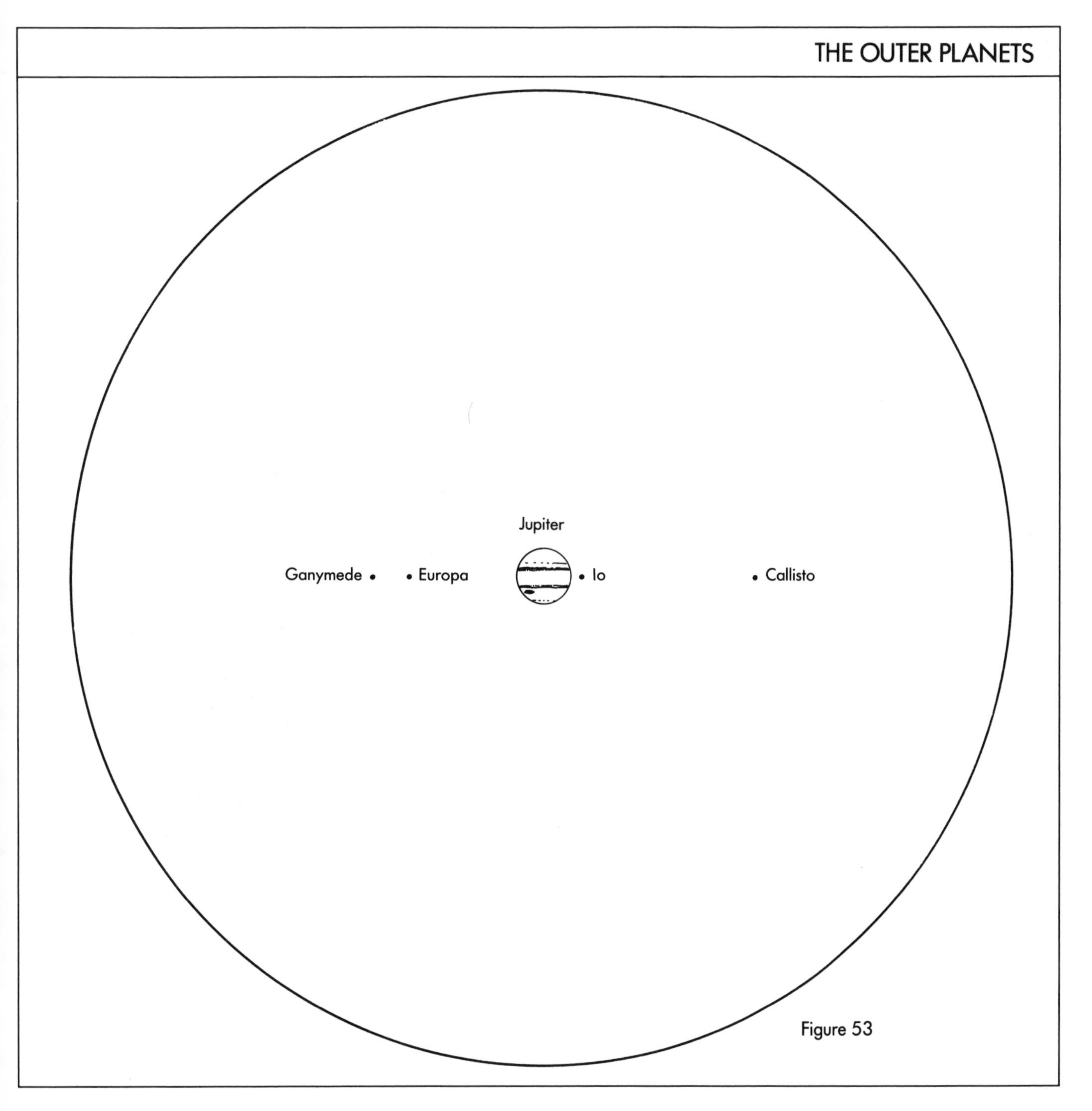

Figure 53

14 JUPITER'S MOONS AND GREAT RED SPOT

LEVEL: Advanced

EQUIPMENT: Same as Project 13, plus a timing source

PROJECT: Observe and draw the moons and Great Red Spot of Jupiter; record the times they appear in different positions; try to determine rotation rate of the planet and to identify the Galilean moons

As you watch Jupiter, it immediately becomes obvious that things on the planet change rather rapidly. From one night to the next, you may see the Great Red Spot, but it will likely be in different positions each night. (See Figures 55 and 56.) Sometimes you may not see it at all, because it is not in our line of sight more than half the time. Jupiter rotates in less than ten hours, however, so on a given night, if you are observing it for several hours, you will be able to see a change in the Great Red Spot's position.

Try recording the time and position of the Great Red Spot every half hour for several hours. Then compare the change in its position with your timing to see if you can determine the time Jupiter takes to make one full rotation. To do your timing, you might try to catch the Great Red Spot in the same position over two consecutive nights. If you can see any other distinct details in Jupiter's cloud bands or any lighter oval spots, try timing the apparent motions of these to help in your study. If you lose sight of a feature you have found but know you are looking at the same part of the planet (which you can determine roughly from where the Great Red Spot is), record that as well.

The moons of Jupiter orbit the planet at different rates. Because our line of sight changes, it is not always possible to determine at first glance which moon is which. Table 13 gives the four largest moons of Jupiter—called the Galilean moons after their discoverer, Galileo—in order going outward from the planet, as well as their rotation periods.

Even easier to see than details in Jupiter's atmosphere are its four largest moons. They resemble stars but always stay fairly close to the planet.

TABLE 13
Revolutions of Jupiter's moons

Moon	Period of revolution
Io	1.769 days
Europa	3.551 days
Ganymede	7.155 days
Callisto	16.689 days

Jupiter was the chief Roman god and is the largest of the planets.

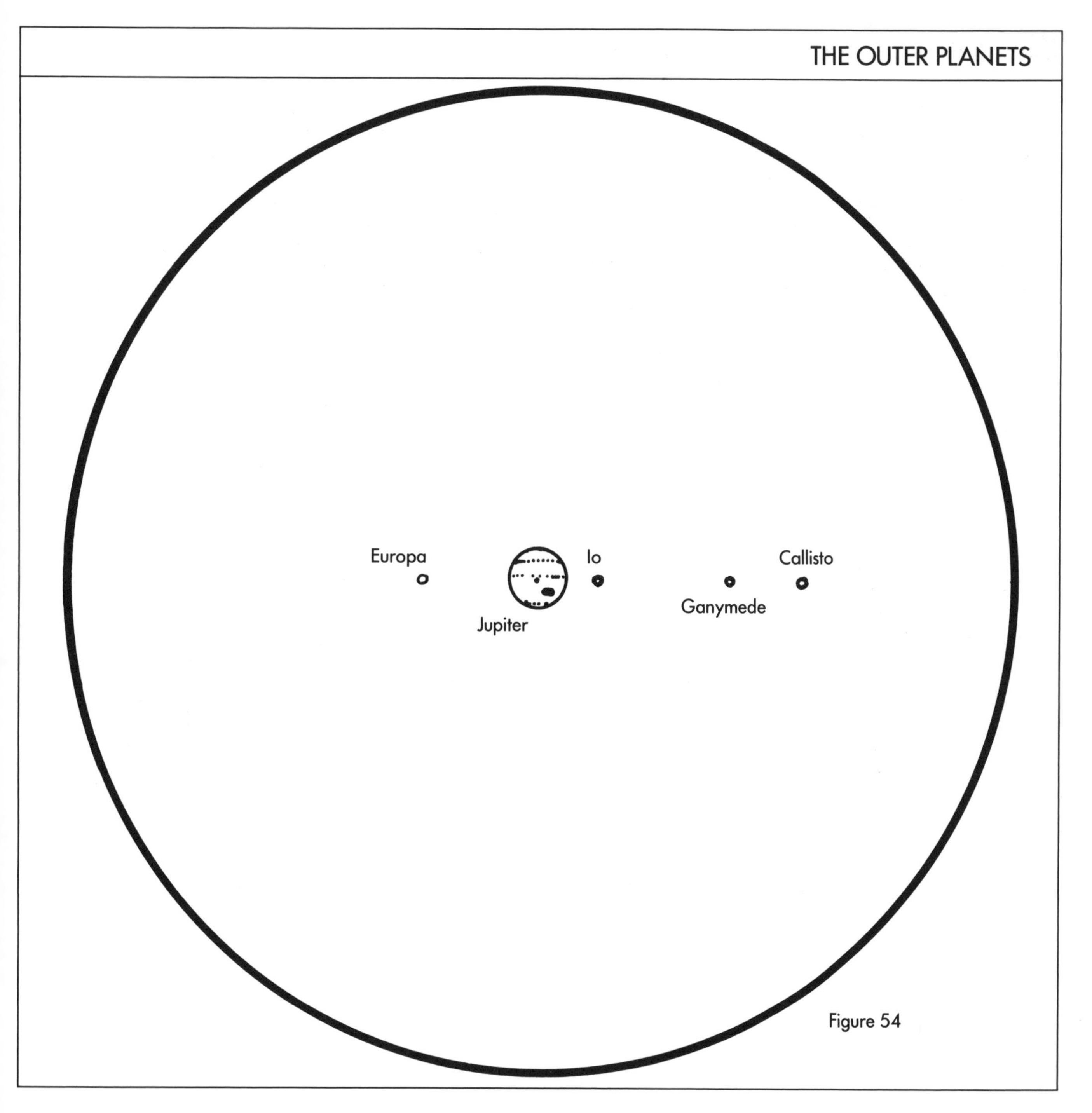

Figure 54

Even the relatively slow Callisto changes from one night to the next. After tracking the moons for several nights to determine, by their speed, which is which, see if you can catch one of them passing behind or in front of the planet. Io is the most likely candidate, as it orbits Jupiter in less than two days. But you might spot one of the others doing it, depending on your patience and timing. If one does appear or disappear at the edge of the planet, time the event for later reference. Try to track Io over several nights and see if you can get a few timings of this sort. Once you have observed and timed Io's orbit, predict times when Io might pass in front of Jupiter. If you can observe at the proper time, you will be able to see an eclipse from a different perspective.

When conditions are right, the shadow of

By watching Jupiter's moons change positions, you can determine which is which. The inner moon, Io, moves fastest, while Callisto exhibits a change much more slowly.

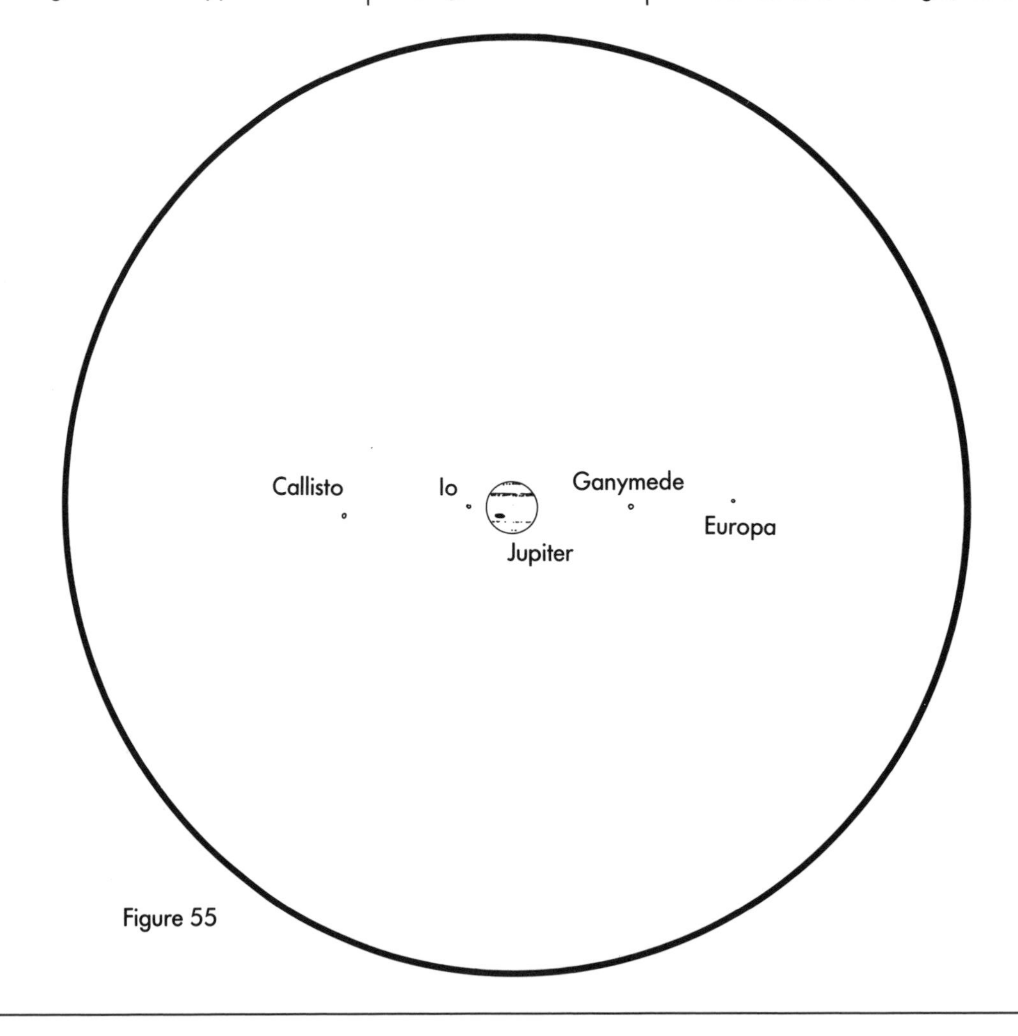

Figure 55

one of the moons will be cast on the upper cloud deck of Jupiter (Figure 56). Time the eclipse parameters if you can. This kind of eclipse is a bit more complicated. Not only may there be a transit of one of the moons in front of the disk, causing the moon to seem to disappear for a while, but the moon will also cast a shadow as it passes by. Once in a great while one of the moons might seem to disappear briefly, then reappear. This is caused by a "mutual eclipse," when the shadow of one moon passes over another moon. There are also even times when the moons can occult each other if the Earth is in the plane of the moons' orbits. These are a few of the many interesting phenomena connected with the moons and features of Jupiter.

If conditions are right, you may be able to see an eclipse on Jupiter's cloud tops. The moon casting the shadow may be difficult to distinguish against the disk of the planet.

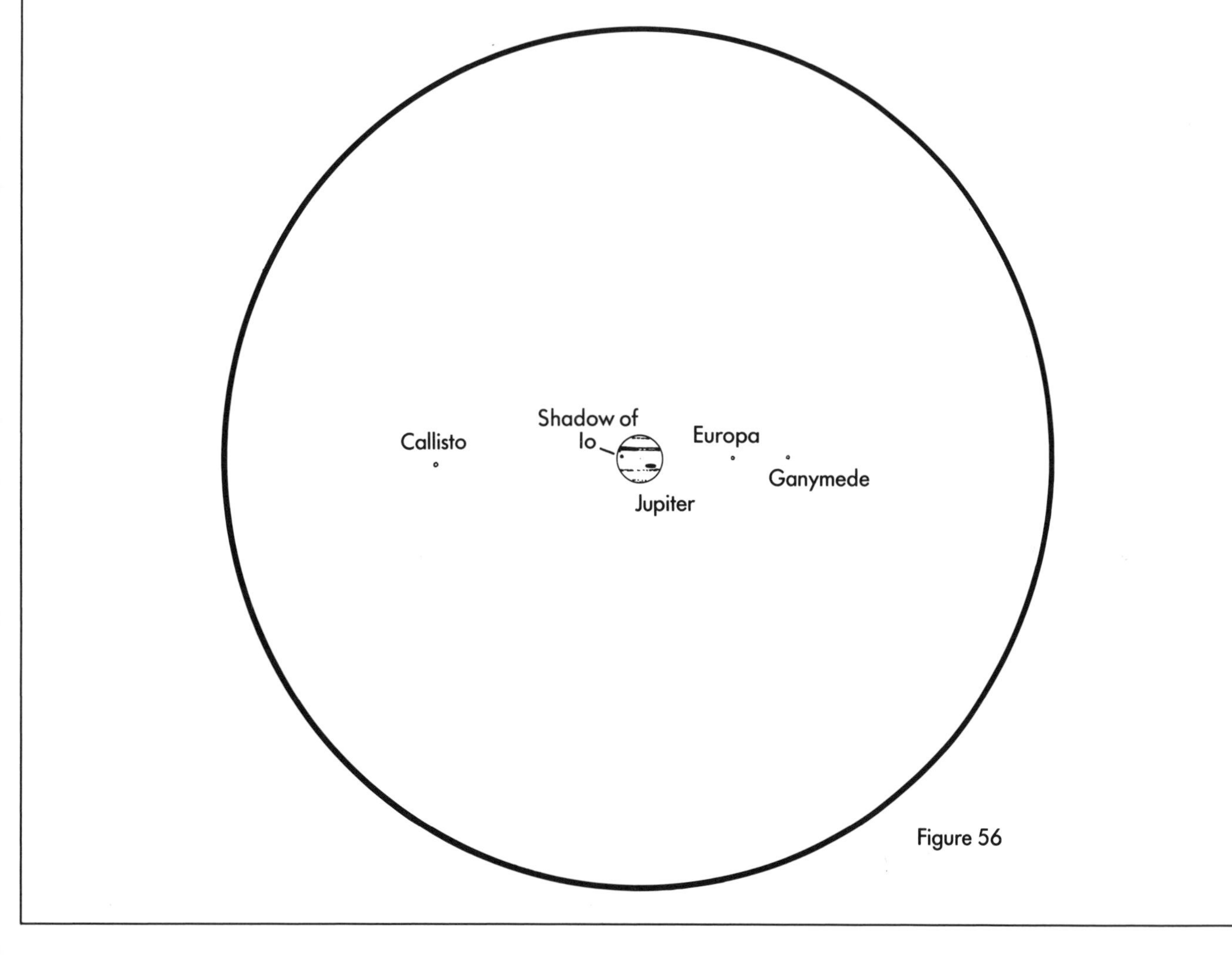

Figure 56

15 SATURN'S RINGS AND MOONS

LEVEL: Intermediate

EQUIPMENT: Same as Project 13, but a spotting scope can be used if a telescope is not available

PROJECT: Observe and draw the rings of and moons of Saturn

Saturn is considerably fainter than Jupiter, but almost the same color. To most observers, it appears to be just a little more yellowish than Jupiter. At best, Saturn is almost twice as far from Earth as its larger neighbor. But because of the extensive and beautiful system of rings that circles Saturn, even powerful binoculars (and definitely a spotting scope) can reveal some details. Telescopes can show the subtle banding in the planet's atmosphere, which at the greater, colder distance of Saturn from the Sun has an obscuring layer of haze. There is nothing as striking in Saturn's atmosphere as the Great Red Spot of Jupiter, but space-probe photographs have revealed banding, turbulence, and spots similar to those on Jupiter. By far Saturn's most impressive feature is its ring system. Although all the gas giant planets have rings, Saturn is the only one with a large, dense system that can be seen from Earth by the amateur astronomer. The ring of Jupiter was first observed from behind the planet by the *Voyager* spacecraft, which was able to capture sunlight backlighting the ring to reveal particles as small as those that make up cigarette smoke. The rings of Uranus and parts of the rings of Neptune have been glimpsed indirectly by watching the light of stars wink on and off as these planets passed nearly in front of them. The starlight was being obscured by thin, dark rings of dust and rock. Circling Saturn is a ring system of rock, dust, and ice with particles that vary in size from a grain of sand or smaller to chunks as large as a small automobile.

Saturn appears fairly small most of the time. You can see it best when it is closest to Earth.

Saturn's ring system varies in density so greatly that it is said to be made up of more than a thousand individual ringlets, which orbit at various distances and interact with the Saturnian magnetic and gravitational fields. The result, as seen from Earth through a small telescope, looks almost like a cream-colored phonograph record with a large hole in the middle. The ring system is usually tilted in relation to our point of view, but sometimes we see the rings edge-on (Figure 59). The rings are divided into several sections, two of which are visible immediately with very little magnification. The large Cassini division—named for the French-Italian astronomer who first observed it—is an apparent gap in the ring system near the outer edge. The outer ring is known as the A ring and the inner one the B ring. Inside the most obvious part of the B ring, nearest the planet is the C ring, also called the "crepe" ring. This one is difficult to observe, because it is translucent. Other thin rings, designated D, E, F and G, have been found as well, but they are extremely difficult to see with amateur telescopes. Try drawing what you can see of the rings and any banding that might be visible on the clouds covering the planet.

Another interesting way to observe the rings of Saturn is to watch stars pass behind them. It is sometimes possible to see stars not only disappear behind the rings but also vary in brightness, wink on and off, and show almost full brightness in the Cassini division. The *Observer's Handbook* and most popular astronomy periodicals note when events such as occultations of brighter stars by planets occur.

Saturn is the planet with the greatest number of known moons. The most widely accepted number of Saturn's moons is twenty-two, but there are some who argue that, below a certain size, a moon should be considered a ring particle. The only moon that is particularly obvious is

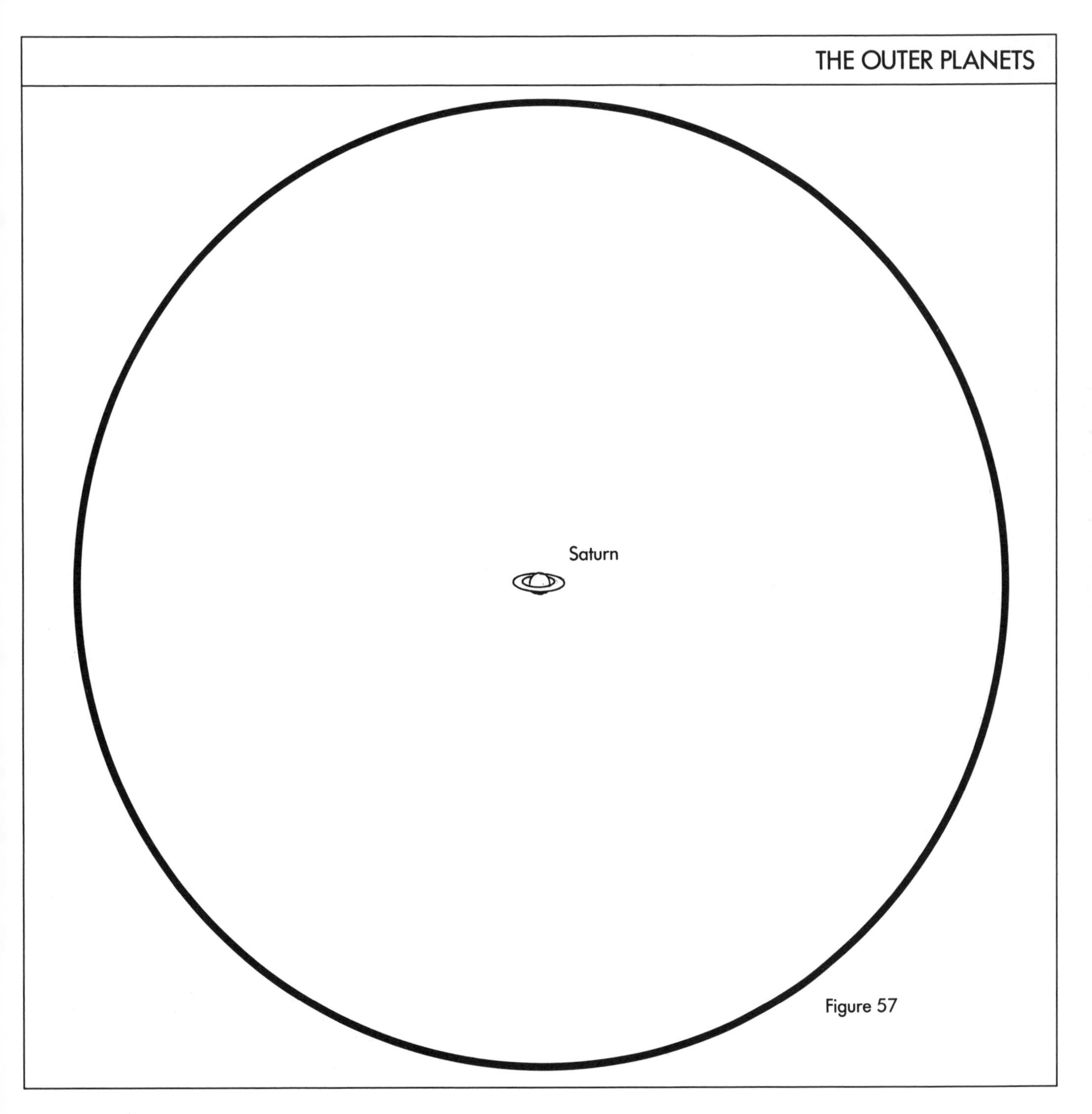

Figure 57

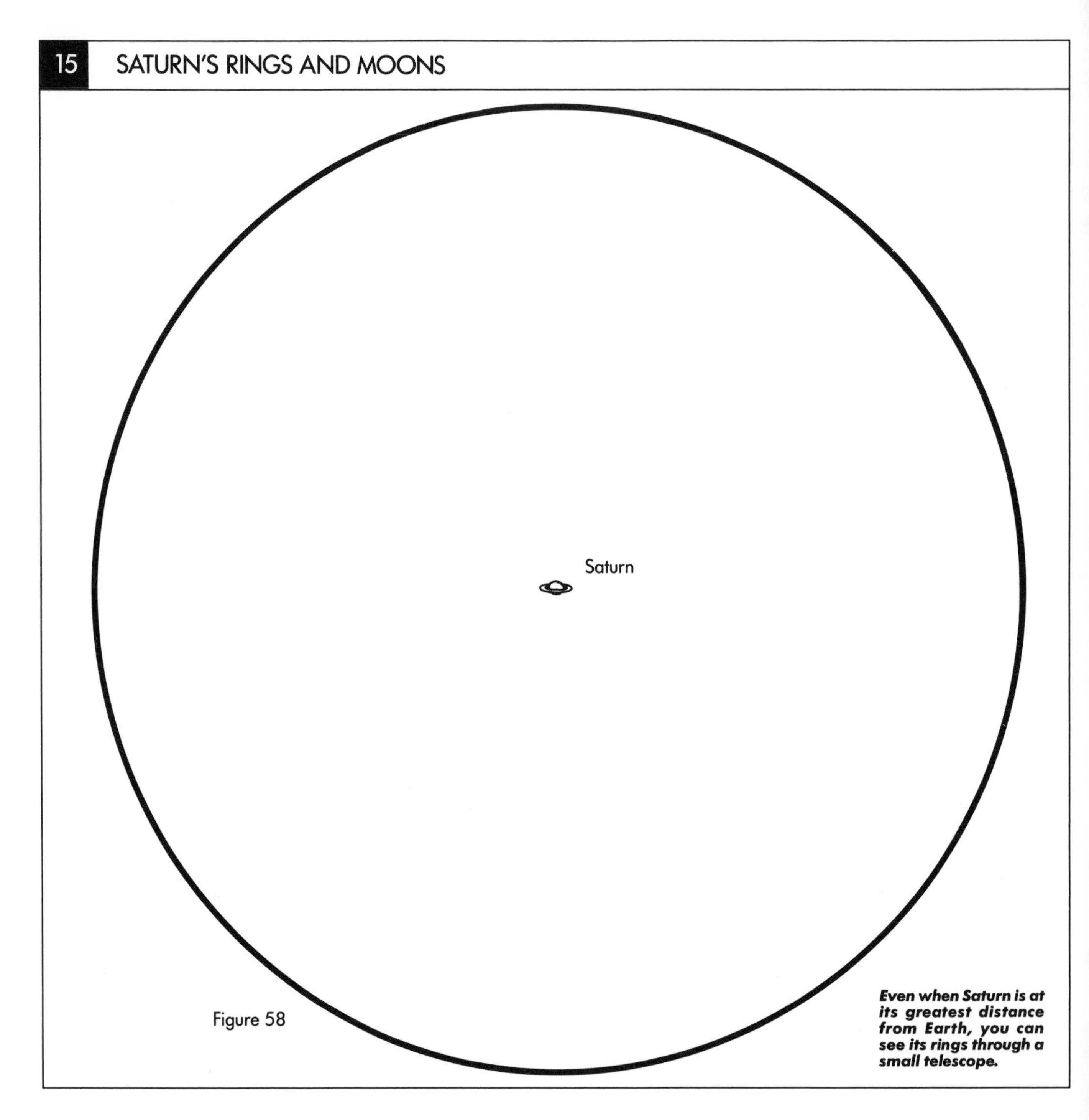

Figure 58

Even when Saturn is at its greatest distance from Earth, you can see its rings through a small telescope.

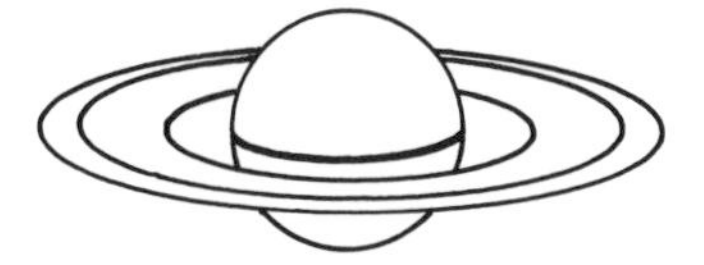
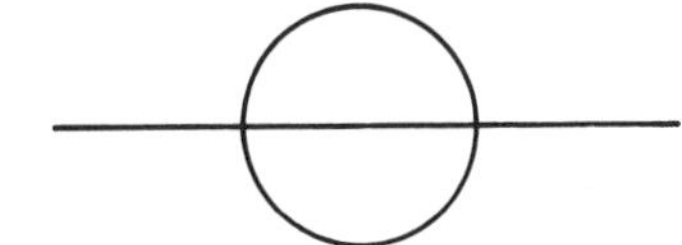

Figure 59

Titan. At about 3,400 miles (5,500 km) in diameter at the cloud tops, this satellite is larger than the planet Mercury. It is covered with a thick nitrogen atmosphere, the major component of our own atmosphere, leading some scientists to believe that it may be one other place in the solar system where living organisms could develop. It appears that some of the other elements necessary for organic compounds can also be found on Titan.

Observing Titan at such a great distance means that it will only be seen as a dot, but it is bright enough that even binoculars could reveal it if it were not so close to the main planet. Look for it at various times. If you think you have found Titan or any other of Saturn's moons (Tethys, Dione, Rhea, Enceladus and Iapetus are also just within the range of small telescopes), record their positions on one of your drawings. Be sure to record the times you observe them as well. Try again another night and see if an object of the same brightness is still there. Titan takes nearly sixteen days to orbit Saturn, so it should not move much from one night to the next. All the other moons mentioned above, except Iapetus, have periods of revolution shorter than five days, so sightings may be a little more difficult to confirm.

Normally we see the rings of Saturn at an angle, which causes them to appear as an open ellipse. At times, Earth passes through the plane of Saturn's rings, and they appear edge-on, which makes them nearly invisible. After Earth passes through the rings' plane, we view them from a new angle, as the other side of the ring system has become visible.

16 PLUTO'S ORBIT

LEVEL: Intermediate

EQUIPMENT: A personal computer and a planetarium simulation program

PROJECT: Locate Pluto at various times over long periods and plot its position against the stars; compare its orbit to the ecliptic to determine the orbit's approximate tilt

Finally come the outermost planets—Uranus, Neptune and Pluto. Even though Uranus *can* just be glimpsed with the unaided eye under ideal visibility conditions, binoculars are needed to reveal its color. It is so far away that it looks like a bluish dot. Even under very high magnification no details can be seen, because the planet's upper cloud deck seems to be covered with a methane haze. Uranus has fifteen known moons, all of which are smaller than our own. These moons, more than 2.5 billion miles (4 billion km) away, cannot be seen even through moderate-size telescopes.

Although harder to find than Uranus, Neptune, at about 3 billion miles (4.8 billion km) away at its closest, actually has a moon large enough to be seen in small telescopes. Again, although photographs dimly show slight features that resemble the two most obvious cloud bands on Jupiter, no features have ever been perceived distinctly against Neptune's bluish cloud tops.

Pluto is definitely outside the range of binoculars and smaller telescopes. It is difficult to observe because it moves so slowly against the star background. On the other hand, Pluto is at its best through 1999, because until then it is in the section of its strangely tilted orbit closest to Earth. In fact, until 1999, Pluto is not the farthest planet from the Sun. For a while, the smallest planet will be in closer than its nearest neighbor, Neptune.

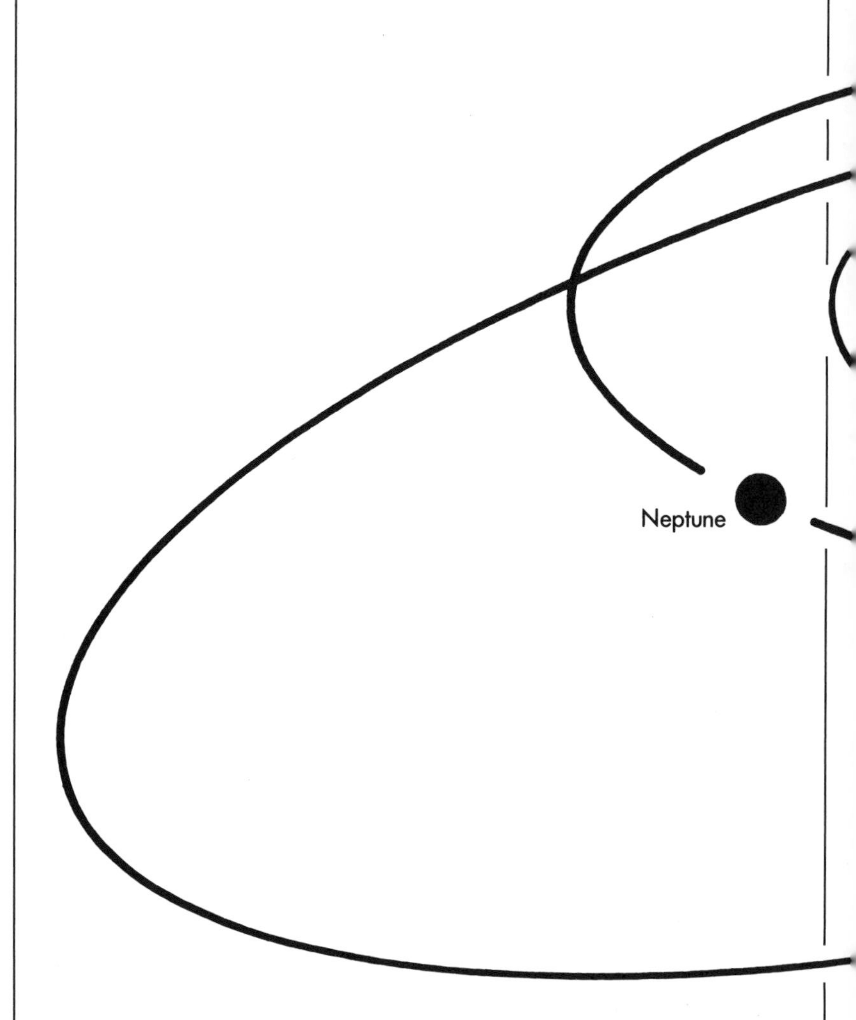

This project can be done using printed references that show the positions of Pluto in the sky at the beginning of each year. Ideally, though, a software package such as Voyager for the Macintosh, SkyGlobe for IBM compatible computers, and Sky Travel for Commodore and Atari computers should be used. The vendor list in

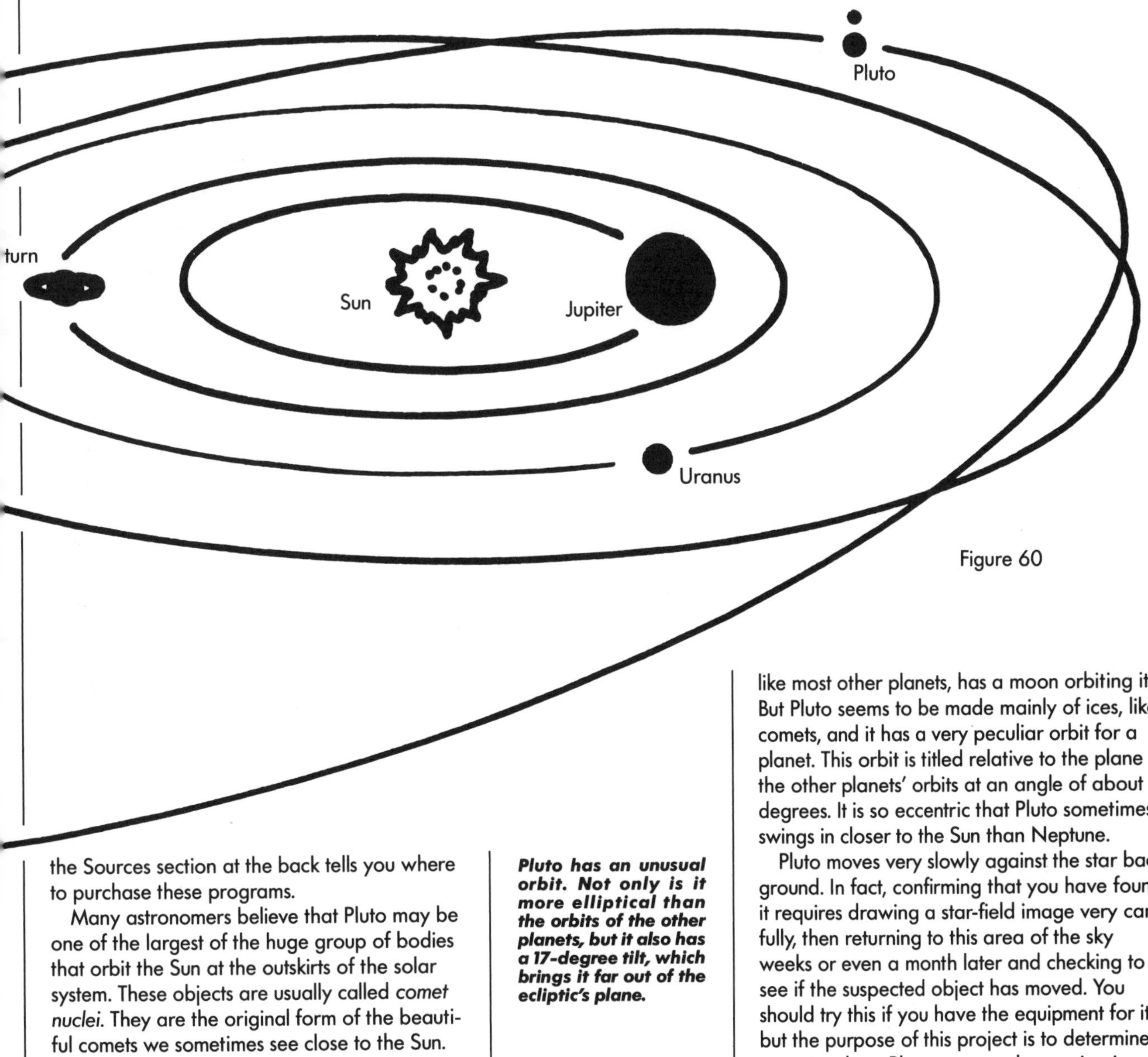

turn
Sun
Jupiter
Pluto
Uranus

Figure 60

the Sources section at the back tells you where to purchase these programs.

Many astronomers believe that Pluto may be one of the largest of the huge group of bodies that orbit the Sun at the outskirts of the solar system. These objects are usually called *comet nuclei*. They are the original form of the beautiful comets we sometimes see close to the Sun. Pluto is large enough to be seen at the inner edge of a suspected shell of such material. It, like most other planets, has a moon orbiting it. But Pluto seems to be made mainly of ices, like comets, and it has a very peculiar orbit for a planet. This orbit is titled relative to the plane of the other planets' orbits at an angle of about 17 degrees. It is so eccentric that Pluto sometimes swings in closer to the Sun than Neptune.

Pluto has an unusual orbit. Not only is it more elliptical than the orbits of the other planets, but it also has a 17-degree tilt, which brings it far out of the ecliptic's plane.

Pluto moves very slowly against the star background. In fact, confirming that you have found it requires drawing a star-field image very carefully, then returning to this area of the sky weeks or even a month later and checking to see if the suspected object has moved. You should try this if you have the equipment for it, but the purpose of this project is to determine over time how Pluto seems to be moving in the sky.

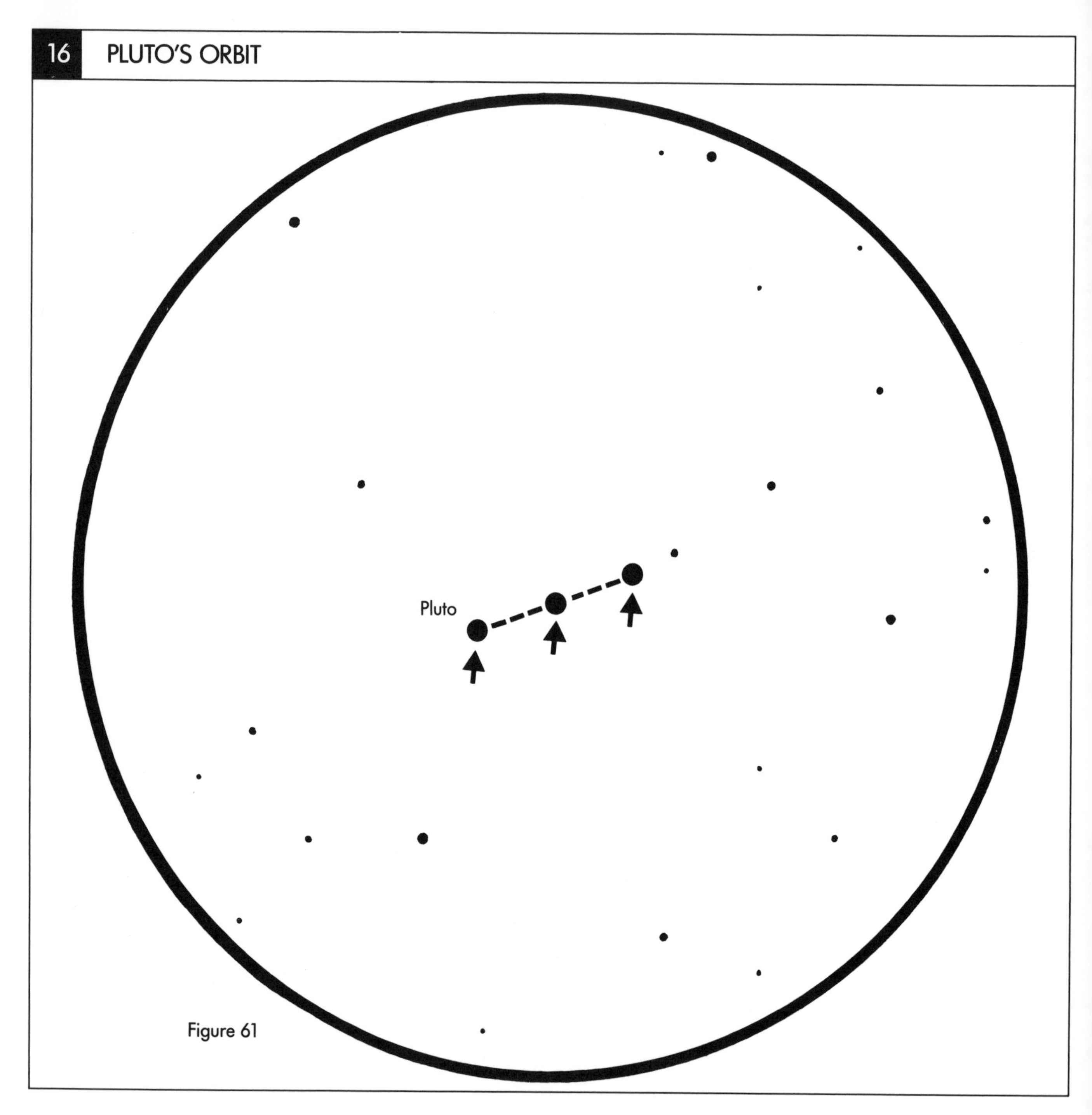

Figure 61

Several software packages (designed for each major type of personal computer) can be used to find any of the planets in the sky. They normally give options of marking the planets with symbols or names to avoid confusion about which planets the user is seeing. With such a software package and one of the star maps provided here, find the next time period when Pluto will be visible in the evening sky. Choose the appropriate star map, then plot Pluto's position on it. Move ahead one year at a time in the sky simulation program, and see how the planet moves. Plot it at each yearly interval for fifty or one hundred years. To complete an entire orbit of Pluto, you could go on for 248 years, but fifty should suffice for this exercise. Once you have your plot, connect the dots on the star map and then measure the angle between this representation of Pluto's orbit and the ecliptic.

You can repeat this project for other planets to find out where they will be in the future, but the interval should be adjusted to months, weeks, or even days, depending on the planet chosen. You will soon begin to discover the effects of retrograde loops, such as the one for Mars mentioned in Project 12, and other interesting things about the apparent motions of the planets.

The only way to identify Pluto is to observe or photograph what you believe to be the planet, at intervals of a week or more, and determine whether the tiny dot has moved the proper distance.

***Several* computer programs are available that create graphic representations of the sky and accurately plot the positions of the planets and stars.**

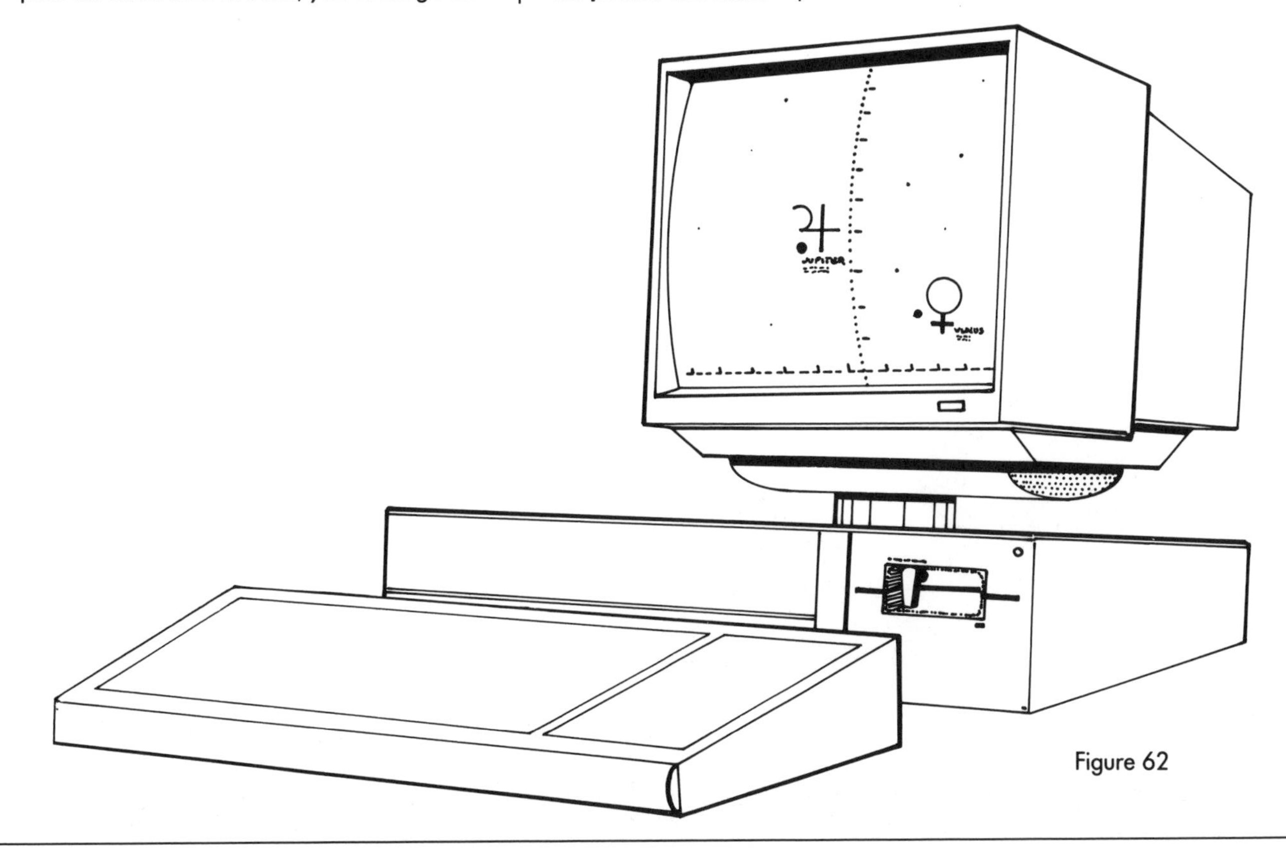

Figure 62

17 METEORS, COMETS, AND AURORAS

LEVEL: Basic

EQUIPMENT: For comets, binoculars, spotting scope or small telescope if searching for new or faint ones—otherwise, only Observation form, pencil, and timing source

PROJECT: Observe some of the more transient phenomena in the sky

The best way to observe auroras and meteors, as well as to conduct general observations of the constellations, is to lie back on a lawn chair and look up. Binocular viewing can be quite comfortable in this position.

Besides the planets, there are several other components of the solar system that are fun and easy to observe. The most popular of these are meteors, comets, and auroras. Easiest to view are meteors, usually seen as faint streaks—"shooting stars"—in the sky. Meteors range in size from tiny particles of dust to huge chunks of rock and metal burning up in our atmosphere. On any clear dark night, anyone can see about six or seven meteors per hour. These drop into the atmosphere at random; they are part of the "stuff" that is out there in the area of space through which the Earth passes in its orbit. The glow they cause in the sky results from a combination of their own burning in the atmosphere (due to friction with the atmosphere) and the glowing of heated air molecules as they move through the air at high speed.

You can observe meteors simply by sitting or lying back somewhere away from direct light and looking up. Try to take in as wide-angle a view of the sky as you can, and simply watch. On any night, you will see sporadic meteors on the average of once every ten minutes, but at

Figure 63

Meteor showers often seem to radiate in toward a focal point in the sky. The shower depicted here has a "radiant" in the constellation Gemini.

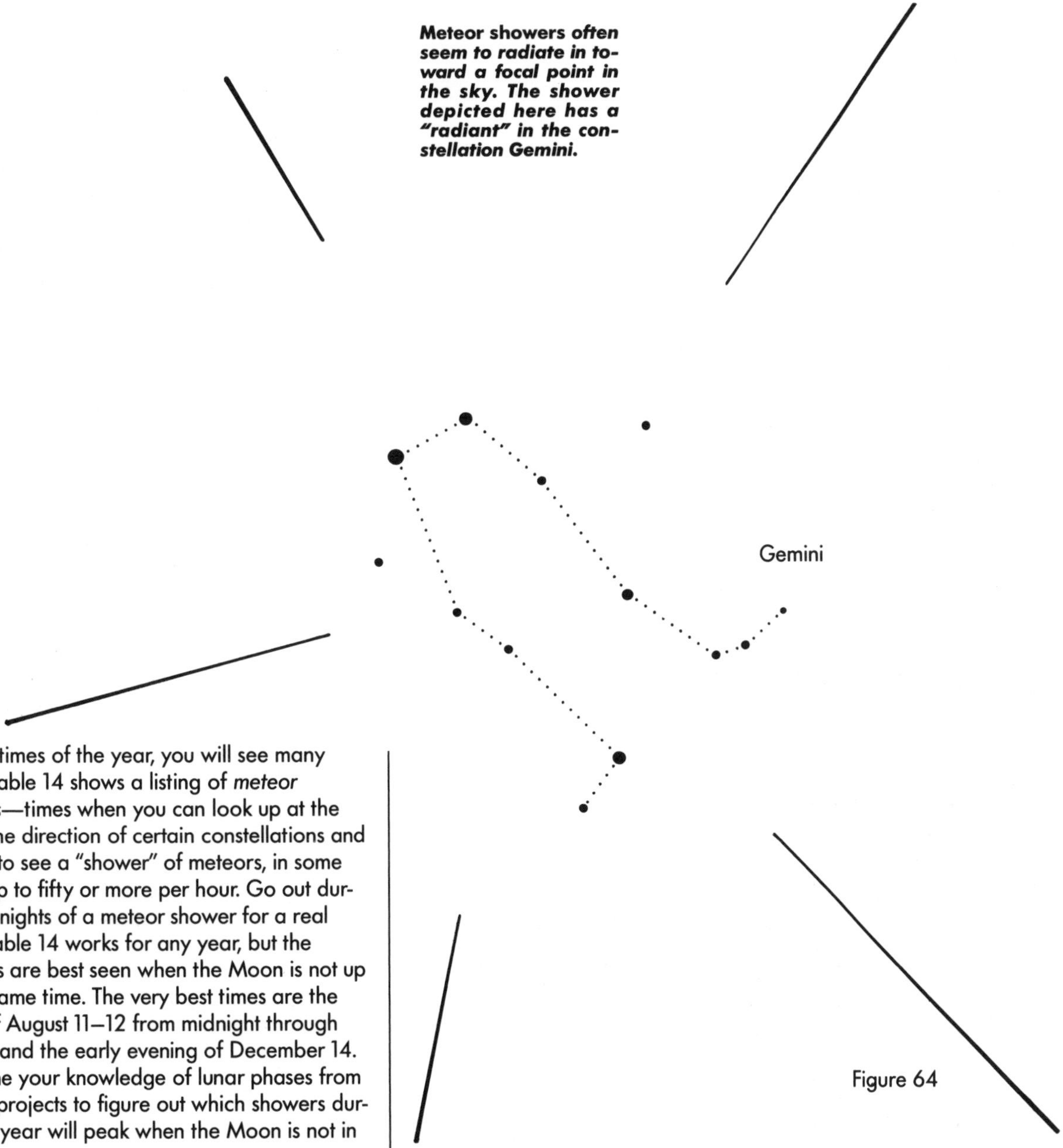

Gemini

Figure 64

certain times of the year, you will see many more. Table 14 shows a listing of *meteor showers*—times when you can look up at the sky in the direction of certain constellations and expect to see a "shower" of meteors, in some cases up to fifty or more per hour. Go out during the nights of a meteor shower for a real treat. Table 14 works for any year, but the showers are best seen when the Moon is not up at the same time. The very best times are the night of August 11–12 from midnight through sunrise and the early evening of December 14. Combine your knowledge of lunar phases from earlier projects to figure out which showers during the year will peak when the Moon is not in the sky.

TABLE 14
Annual Meteor Showers

Meteor shower	Date	Constellation	Number of meteors per hour
Quadrantids	Jan 3	Hercules	40
Lyrids	Apr 22	Lyra	15
S. Aquarids	Jul 28	Aquarius	20
Persieds	Aug 12	Perseus	50
Orionids	Oct 21	Orion	25
Geminids	Dec 14	Gemini	50

Meteor showers are caused by the second phenomenon to observe in this project—the elusive, beautiful comet. The Aquarid (May) and Orionid (October) meteor showers are caused by the Earth passing close to the orbit of the famous comet known as Comet Halley. Comets are chunks of rock and ice that orbit the Sun. It is believed that there are billions of such chunks in the outer solar system, beyond the orbit of Pluto. Gravitational effects cause some of these comet nuclei to be pulled into the inner solar system. Comets usually follow very elliptical paths that bring them in close to the Sun for only a short part of their orbits. As they approach the Sun, they begin to heat up, and gases and dust are released to form the beautiful tails that trail bright comets. Dust and chunks of rock are left behind that can fall into our atmosphere as meteor showers when the Earth passes through the area.

Comets often appear only as faint smudges of light, because only the largest nuclei that move close enough to the Sun form long tails like that for which Comet Halley is famous. Even Halley, when it reappeared in 1985–1986, was not in a good position for viewing from Earth. Since the appearance of bright comets is largely unpredictable, the only "best" time to look for them is when you hear about them in news reports or when your local planetarium or astronomy club announces a comet's arrival on its sky information line. Otherwise, as you observe the sky always keep an eye out for unusual fuzzy patches. If you see one, carefully note its position relative to the stars, and check to find out if it is one of the many galaxies or nebulae shown, say, on the star map in the center of *Sky and Telescope* magazine. If not, it might be a comet. The staff at a local planetarium, college astronomy department, or your nearest astronomy club should be able to help you determine which it is for sure, *if* you record a good position for it on your star map.

One of the most beautiful of all sky phenomena takes us back to the Earth's atmosphere. At least as unpredictable as bright comets are the auroras. Called northern and southern lights (depending on which hemisphere you are in), auroras are caused by particles that are thrown out from the Sun by huge explosions called solar flares. Auroras are predictable to the degree that when a large solar flare is observed, an auroral display can be expected a few days later. But observing conditions, your latitude, and many other factors can affect whether or not you can see them.

Most often observed in high northern or southern latitudes, auroras have at times been seen as far south as Cuba, so they are worth looking for unless you are close to the equator. As with comets, listen for reports on when they might appear, and, with meteors, to see them simply go out and look up. Record your impressions of the shape (curtain-like, streaks, bursts) and color and the time you see them. If you miss a display, don't give up. Each time auroras are predicted gives you another chance.

It is difficult to predict when comets will appear, because they often have greatly elongated elliptical orbits around the Sun. There are perhaps millions of comets in the solar system, but they remain invisible until they get close enough to the Sun to form their tails.

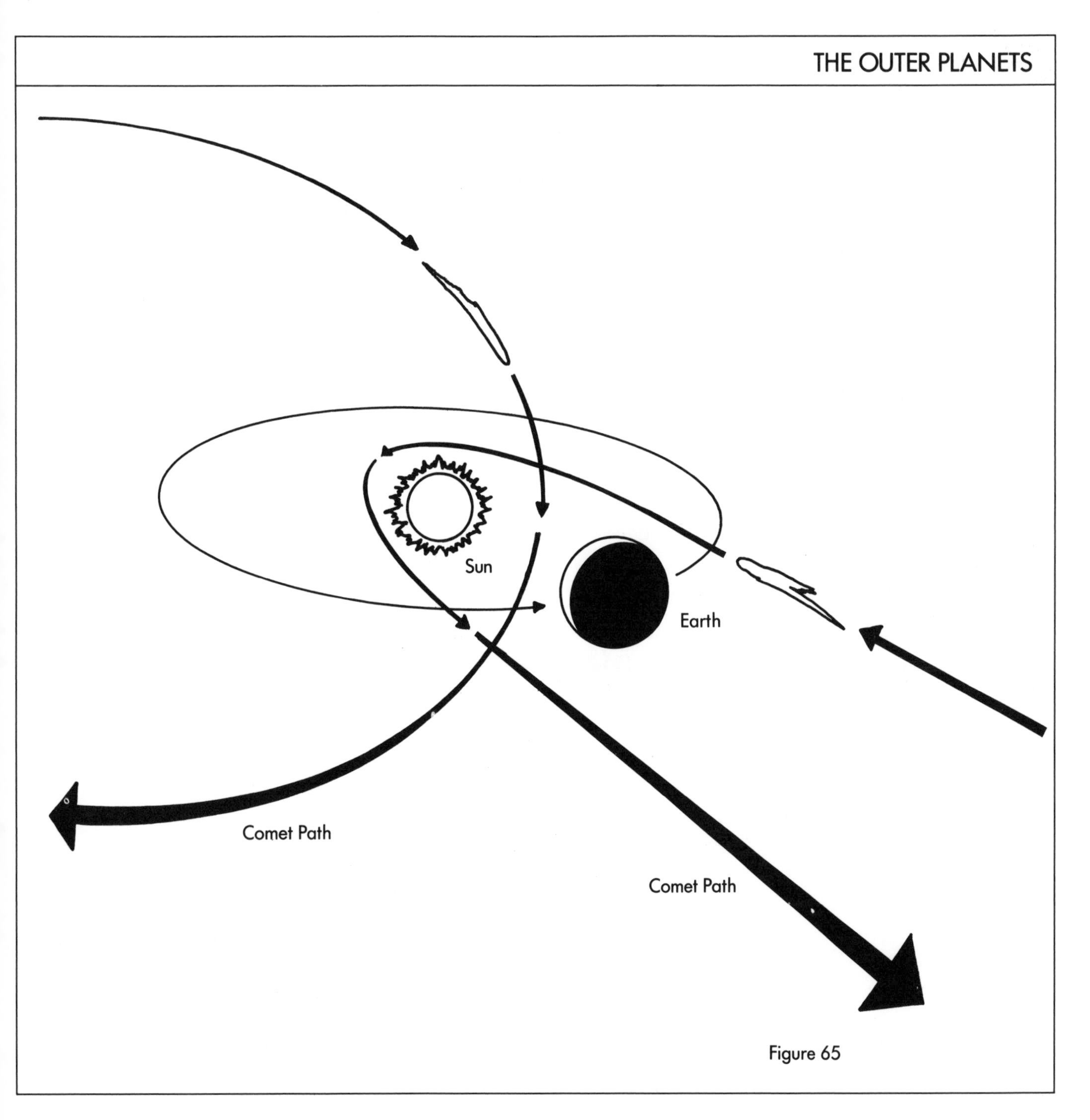

Figure 65

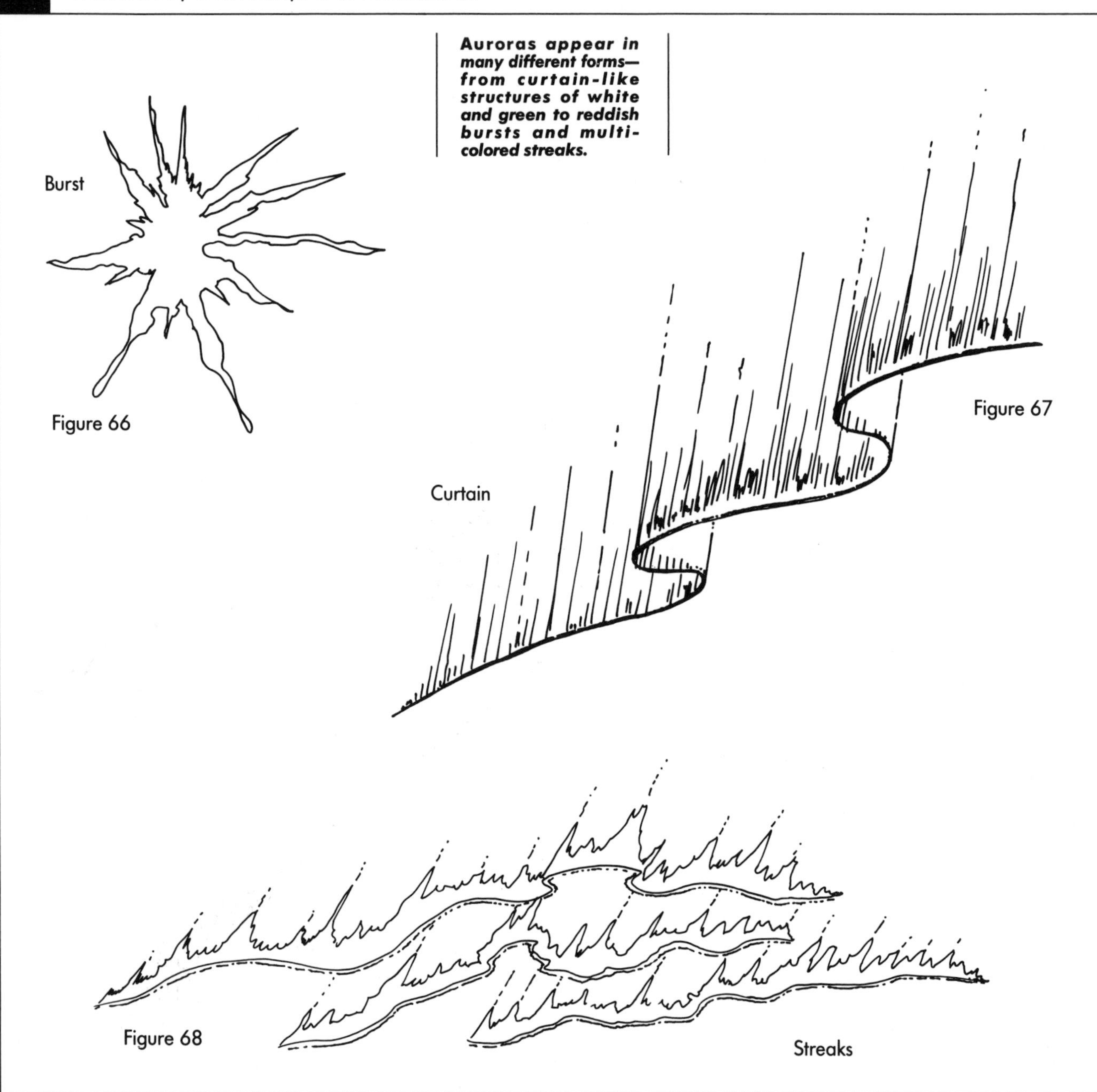

Figure 66

Figure 67

Figure 68

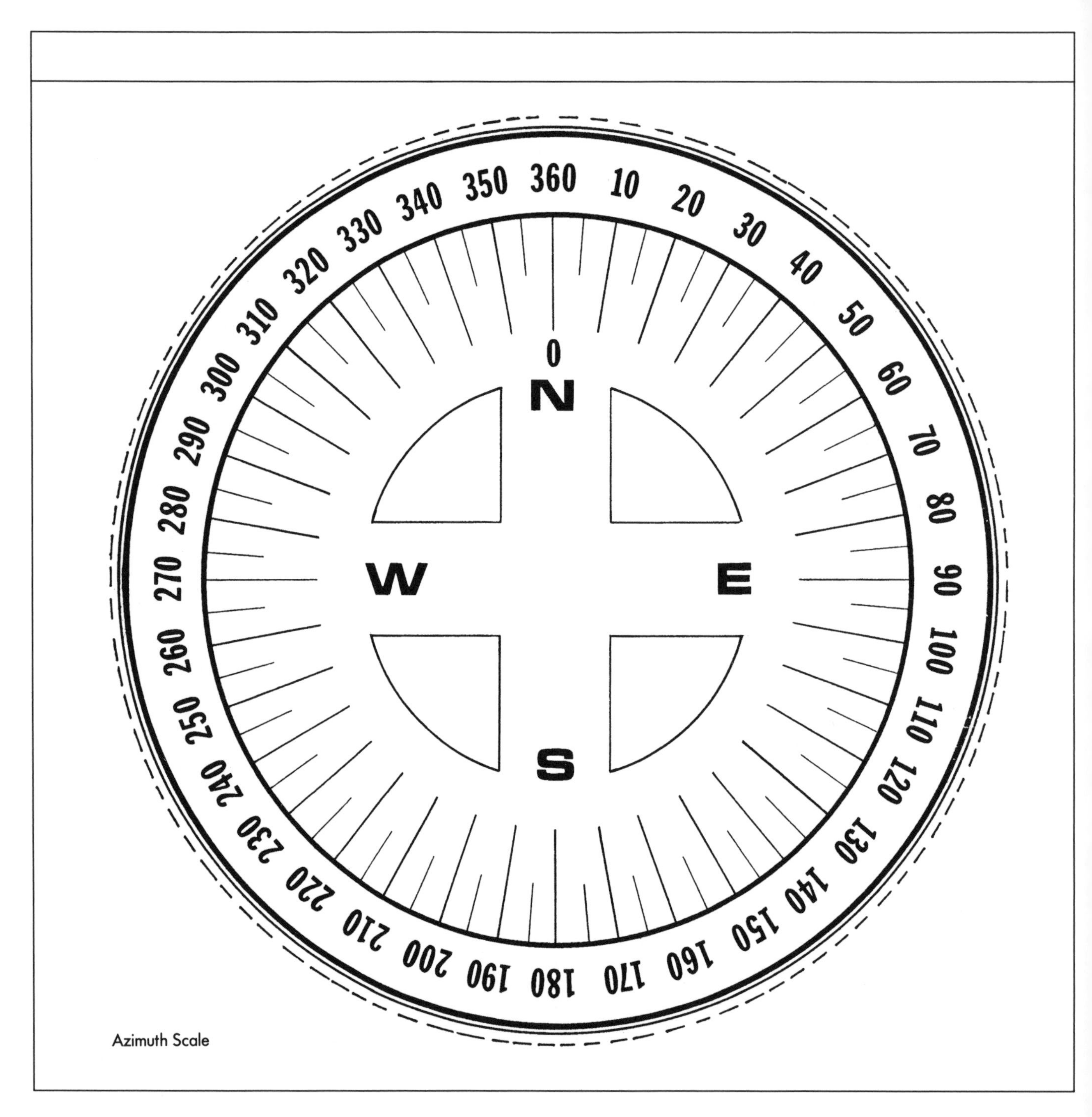

Azimuth Scale

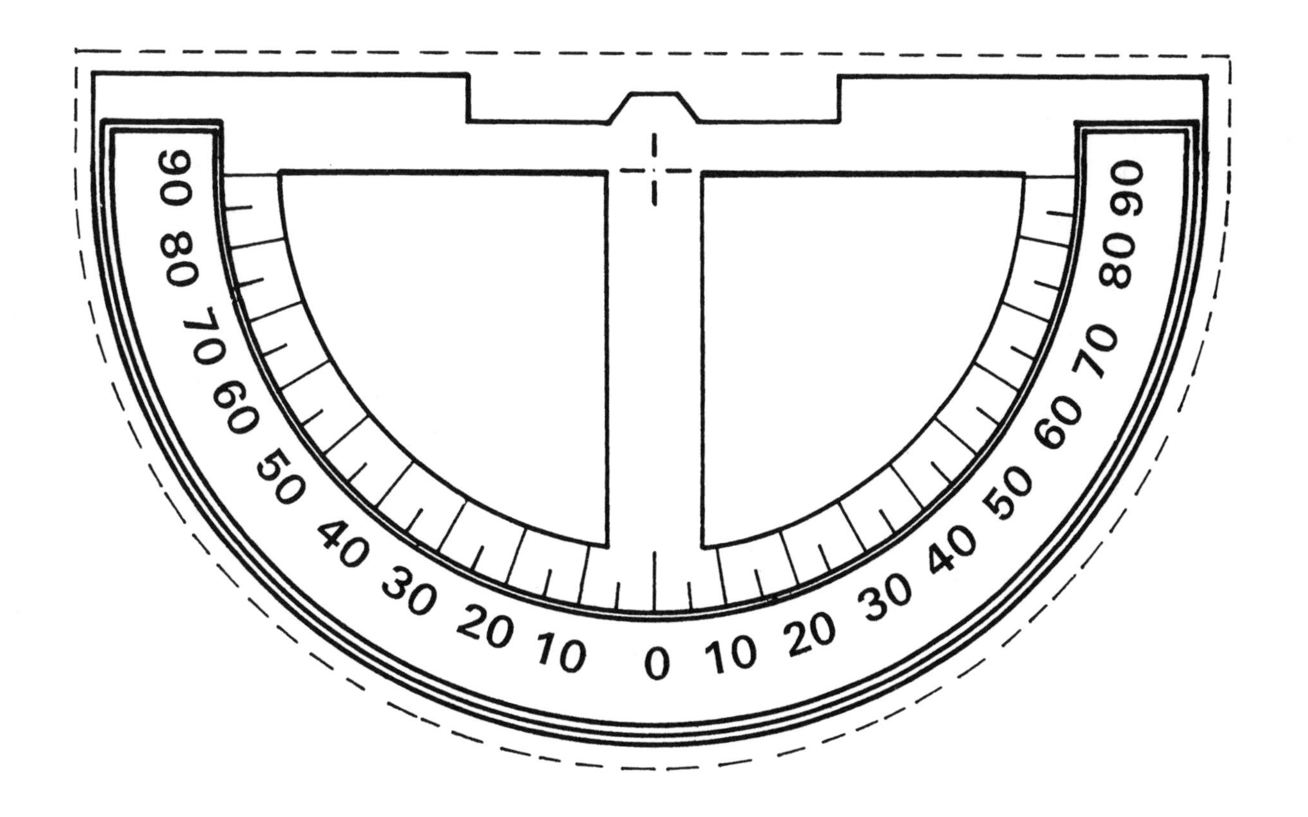

Altitude Scale

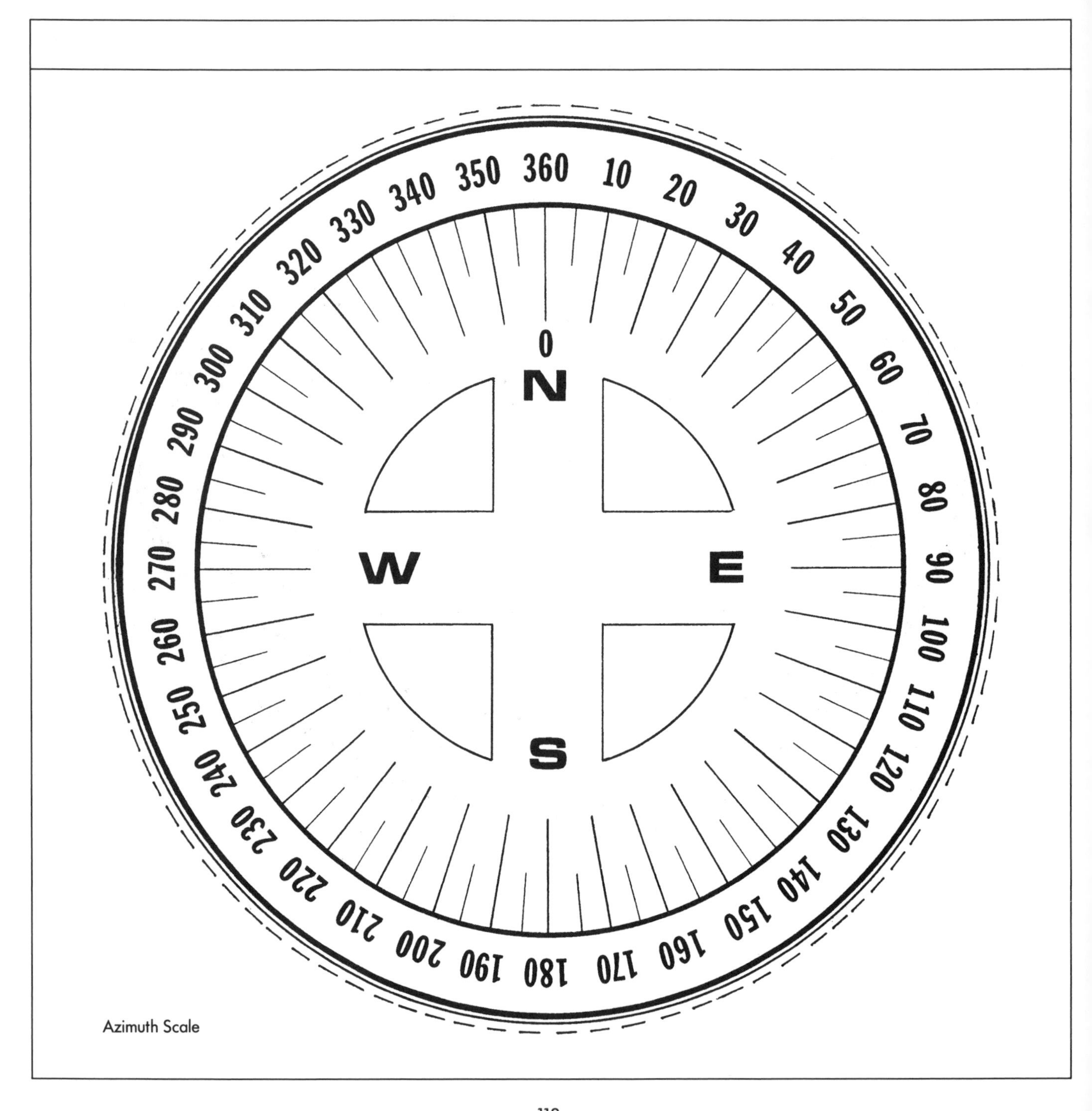

Azimuth Scale

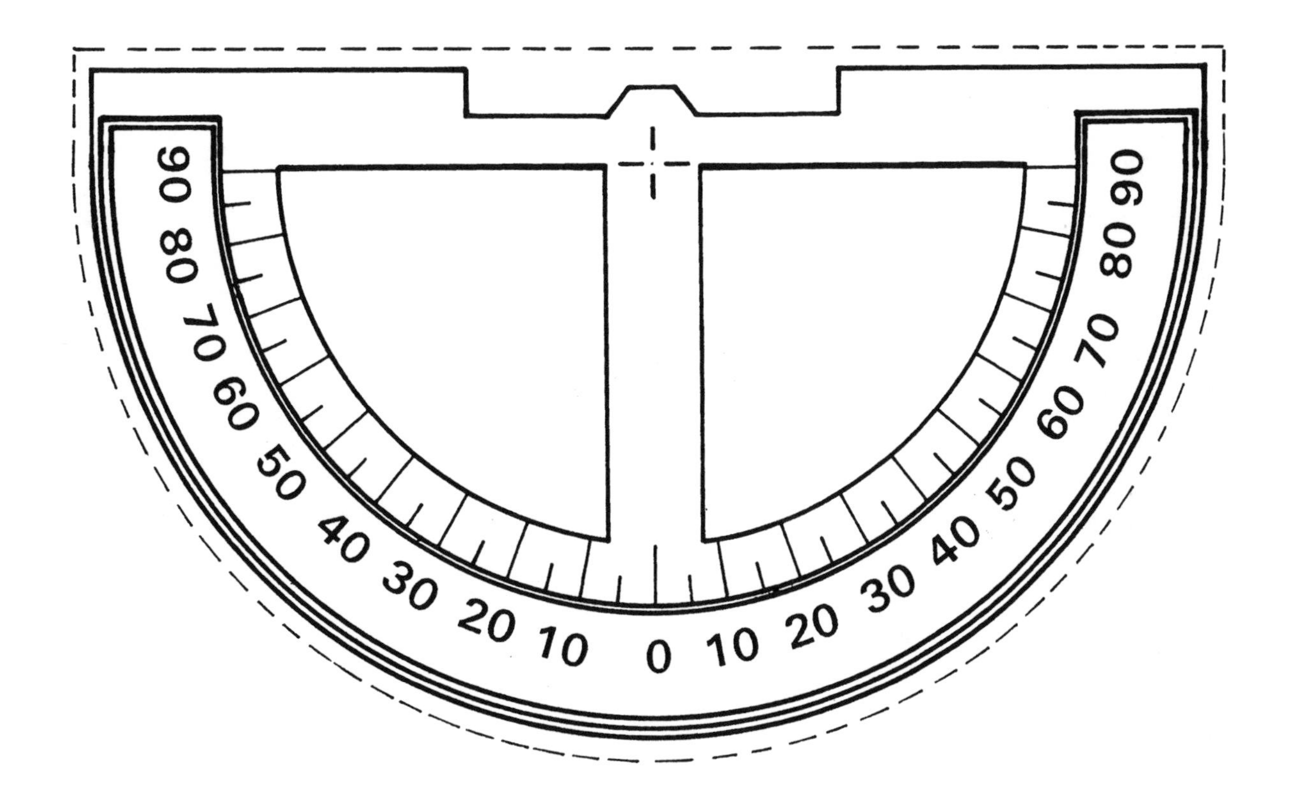

Altitude Scale

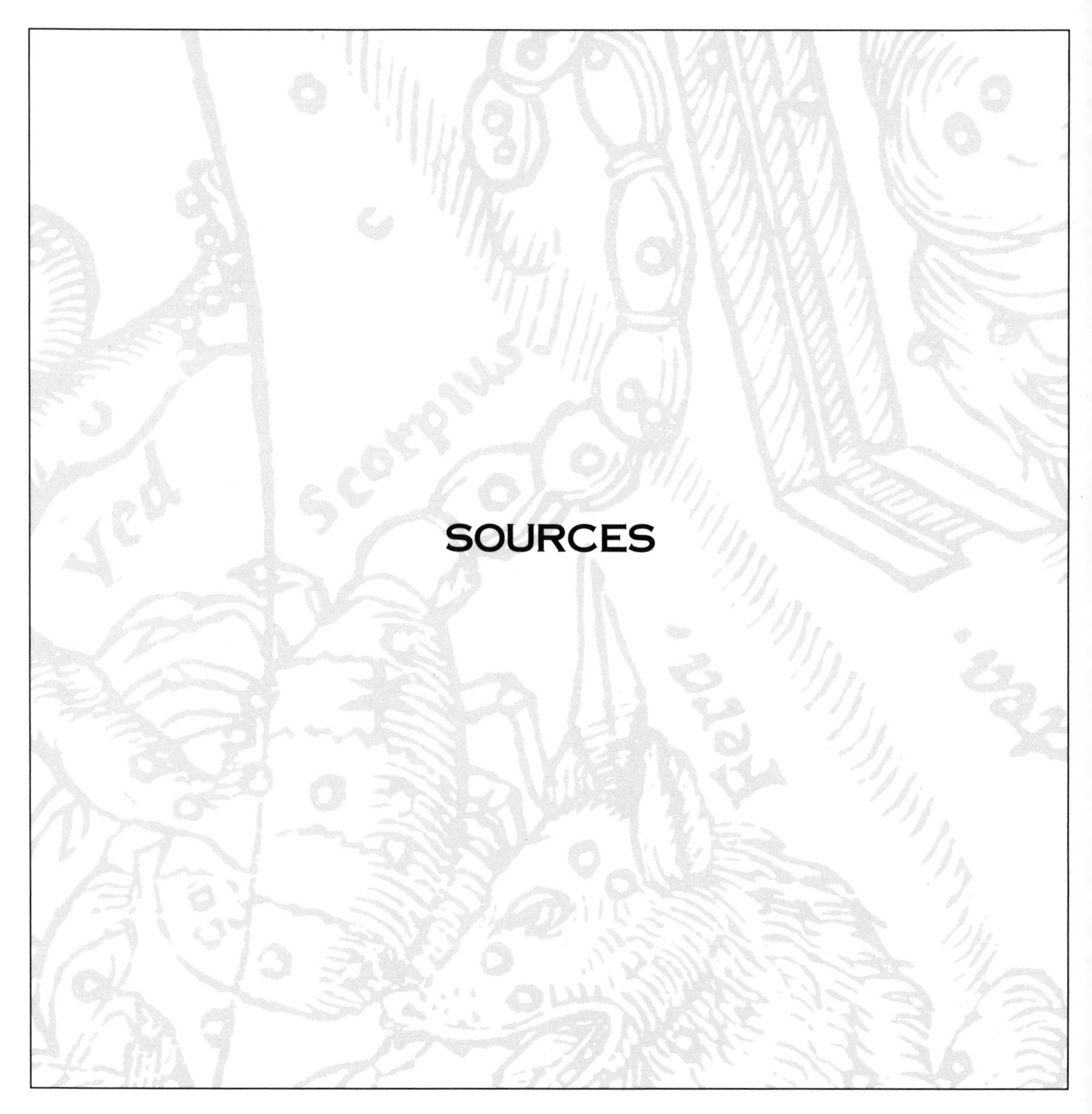

SOURCES

VENDORS OF ASTRONOMICAL EQUIPMENT (ALSO SEE ADS IN *ASTRONOMY* AND *SKY AND TELESCOPE*)

Celestron International
2835 Columbia Street
Torrance, CA 90503

Commodore Electronics, Ltd.
(For information on Sky Travel)
1200 Wilson Dr.
West Chester, PA 19380

Edmund Scientific Company
Department 12B1
N964 Edscorp Building
Barrington, NJ 08007

Meade Instruments Corporation
1675 Toronto Way
Costa Mesa, CA 92626

Sun Spotter II (for viewing the Sun without a telescope):
RD1 Box 160
Hawley, PA 18428
(717) 685-7033

Swift Instruments, Inc.
952 Dorchester Avenue
Boston, MA 02125

Eclipse filters for telescopes are available from telescope manufacturers above.

SOURCES OF SKY SIMULATION COMPUTER PROGRAMS

ARC Science Simulations (Dance of the Planets for IBM PC/AT)
P.O. Box 1955
Loveland, CO 80539

Carina Software (Voyager for Macintosh computers)
830 Williams Street
San Leandro, CA 94577

KlassM Software (SkyGlobe)
284 142nd Avenue
Caledonia, MI 49316

Zephyr Services (several programs)
1900 Murray Avenue, Department B
Pittsburgh, PA 15217

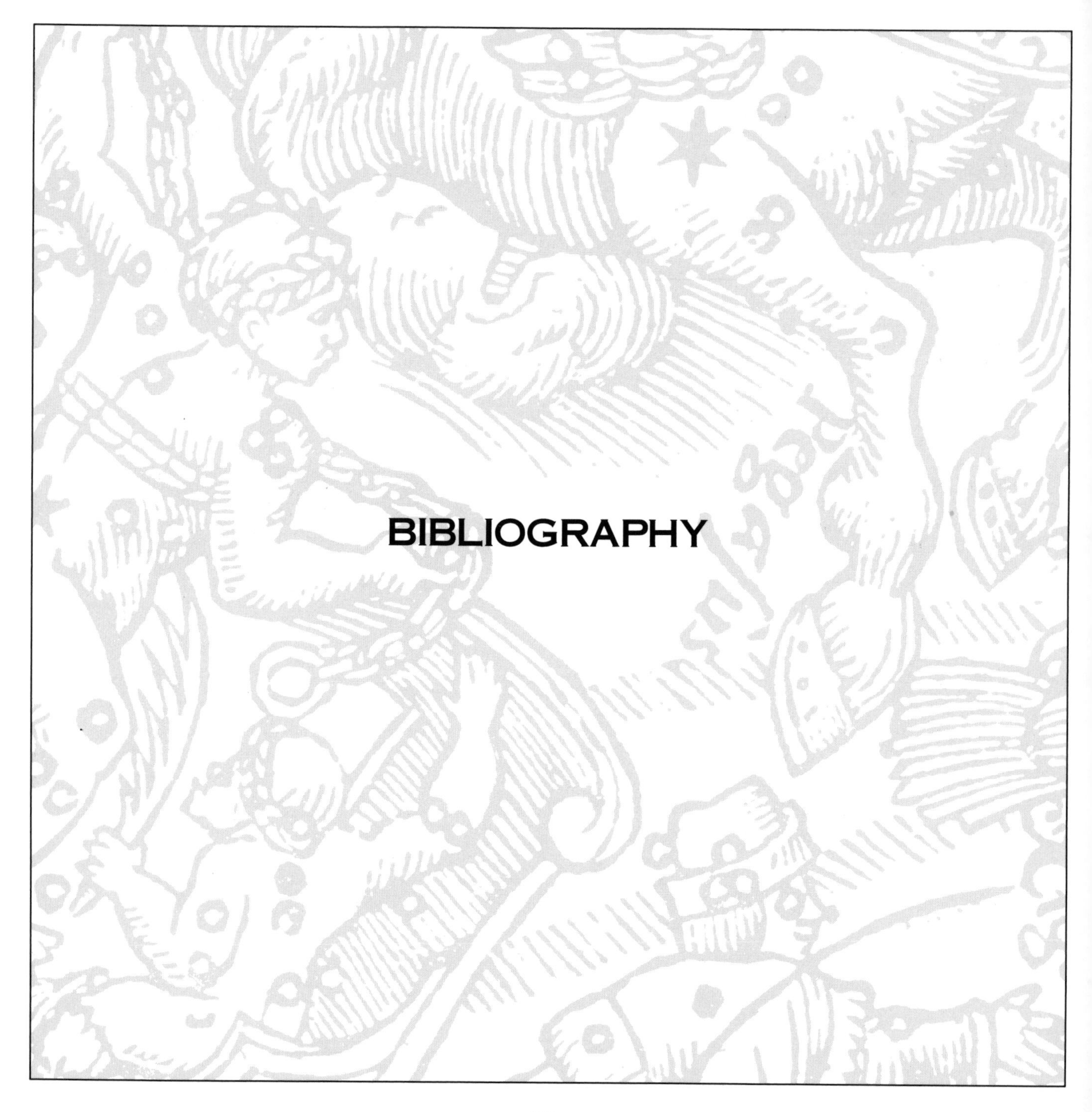

BIBLIOGRAPHY

Astronomical Almanac (annual)
Superintendent of Documents
U.S. Government Printing Office
Washington, D.C. 20402
(Available in some major libraries, especially libraries of universities with astronomy programs.)

The Journal
Association of Lunar and Planetary Observers (ALPO)
P.O. Box 143
Heber Springs, AK 72543

Astronomy Magazine (monthly)
Kalmbach Publishing Company
21027 Crossroads Circle
P.O. Box 1612
Waukesha, WI 53187
(Available at most major bookstores and libraries.)

IAU Lunar Atlas, 1935
Smithsonian Astrophysical Observatory
60 Garden Street
Cambridge, MA 02138

The Moon Observer's Handbook
Fred W. Price
Cambridge University Press
32 East 57th Street
New York, NY 10022
(Available in many public libraries.)

Observer's Handbook (annual)
The Royal Astronomical Society of Canada
136 Dupont Street
Toronto, Ontario M5R 1V2 Canada

Sky Atlas 2000
Sky Publishing Corporation
49 Bay State Road
Cambridge, MA 02238-1290
(Available in some major bookstores.)

Sky and Telescope Magazine (monthly)
Sky Publishing Corporation
49 Bay State Road
Cambridge, MA 02138-1290
(Available in some major bookstores and libraries.)

GLOSSARY

Altitude. The angle measured, in degrees, upward from the horizon to an object in the sky. The altitude of a perfect horizon is 0 degrees. The zenith (see below) is at 90 degrees. Halfway up in the sky is 45 degrees.

Auroras. Luminous streamers visible in the upper atmosphere in arctic and subarctic regions. Caused by particles from the Sun reacting with molecules in the atmosphere, making them glow. The aurora borealis is also called the northern lights; the aurora australis is also known as the southern lights.

Azimuth. The angle measure, in degrees from true north, toward the east, to the point on the horizon that is directly below an object in the sky. North is at azimuth 0 degrees, east at 90, south at 180 and west at 270 degrees.

Celestial pole. The point in the sky that is exactly above one of the poles of the Earth. Established by extending an imaginary line from our North or South Pole into the sky.

Celestial equator. A projection of the Earth's equator onto the celestial sphere. It is a great circle on which all points are 90 degrees from the celestial poles.

Comet. A small chunk of rock and ices that orbits the Sun in a highly eccentric path. Comets become visible as they approach the Sun. Gases evaporate and dust is released to form the coma, or head, of the comet, and a tail that can extend for more than 100 million miles. The light we see from comets is sunlight reflected by this material.

Crater. A depression (usually circular) on a planet or satellite. Craters are formed by the impact of meteorites on the surface of the body.

Declination. The angular distance of a position or object in the sky measured north or south of the celestial equator. Negative numbers are used south of the equator, positive to the north. The celestial equator is at declination 0. The north celestial pole is at 90 degrees.

Ecliptic. A great circle on the celestial sphere that marks the apparent path of the Sun against the stars.

Foreshortening. An effect that occurs when looking at an object at an oblique angle rather than seeing it from directly above, which causes the object to look smaller along the line of sight. A crater on the moon is circular and is seen as such when viewed from directly overhead, but appears elliptical when viewed from any other angle.

Greatest elongation. The point at which the apparent angle between the Sun and a planet, as seen from Earth, is at its greatest.

Horizon. The lowest point in the sky that can be seen by an observer. The ideal horizon is a plane exactly perpendicular to the zenith (see below).

Kilometer. A unit of length measure used by scientists (and by people in most countries of the world); a kilometer (abbrev., km) is 1,000 meters, or about .62137 miles.

Mare. The Latin name for a sea-like lunar feature; plural, maria.

Meteoroid, meteor, meteorite. A meteoroid is a chunk of rock or dust, part of the debris that floats in outer space. A meteor is the bright streak of light that is visible when a meteoroid passes through the Earth's atmosphere and is heated by friction between it and air molecules. A meteorite is a meteoroid that has survived the trip all the way to the surface of Earth.

Occultation. The passage of an apparently larger body in front of a seemingly smaller body in the sky, for example, the Moon seeming to pass in front of stars or planets.

Polaris. The North Star, located at the end of the handle of the Little Dipper. Known by astronomers as Alpha Ursa Minoris.

Right ascension. The angular distance measured from west to east along the celestial equator. Right ascension can be thought of as being like an extension of longitude out onto the celestial sphere. However, it is measured in hours (1 hour = 15 degrees) and is fixed with respect to the sky rather than the Earth.

Rille. A crevasse or channel on the lunar surface. Some rilles appear to have been cut by liquid action, probably volcanic flows in the ancient lunar past.

Speed of Light/Light year. The speed which light travels has been estimated at 186,272 miles per second. A light year is the distance light travels in one year's time, or just under 6 trillion miles (more than 9.6 trillion km).

WWV. The call letters of a broadcast service that provides accurate time signals over shortwave radio. These signals originate at the National Bureau of Standards.

Zenith. The point in the sky that is directly overhead for a given observer. Each observer on Earth has a different zenith.

Zodiac. An imaginary band in the sky extending about 10 degrees on either side of the ecliptic, roughly shaped by twelve of the constellations through which the Sun seems to travel as Earth orbits it. The Sun, Moon and all the planets except (at times) Pluto, can be seen against the constellations of the zodiac.

INDEX

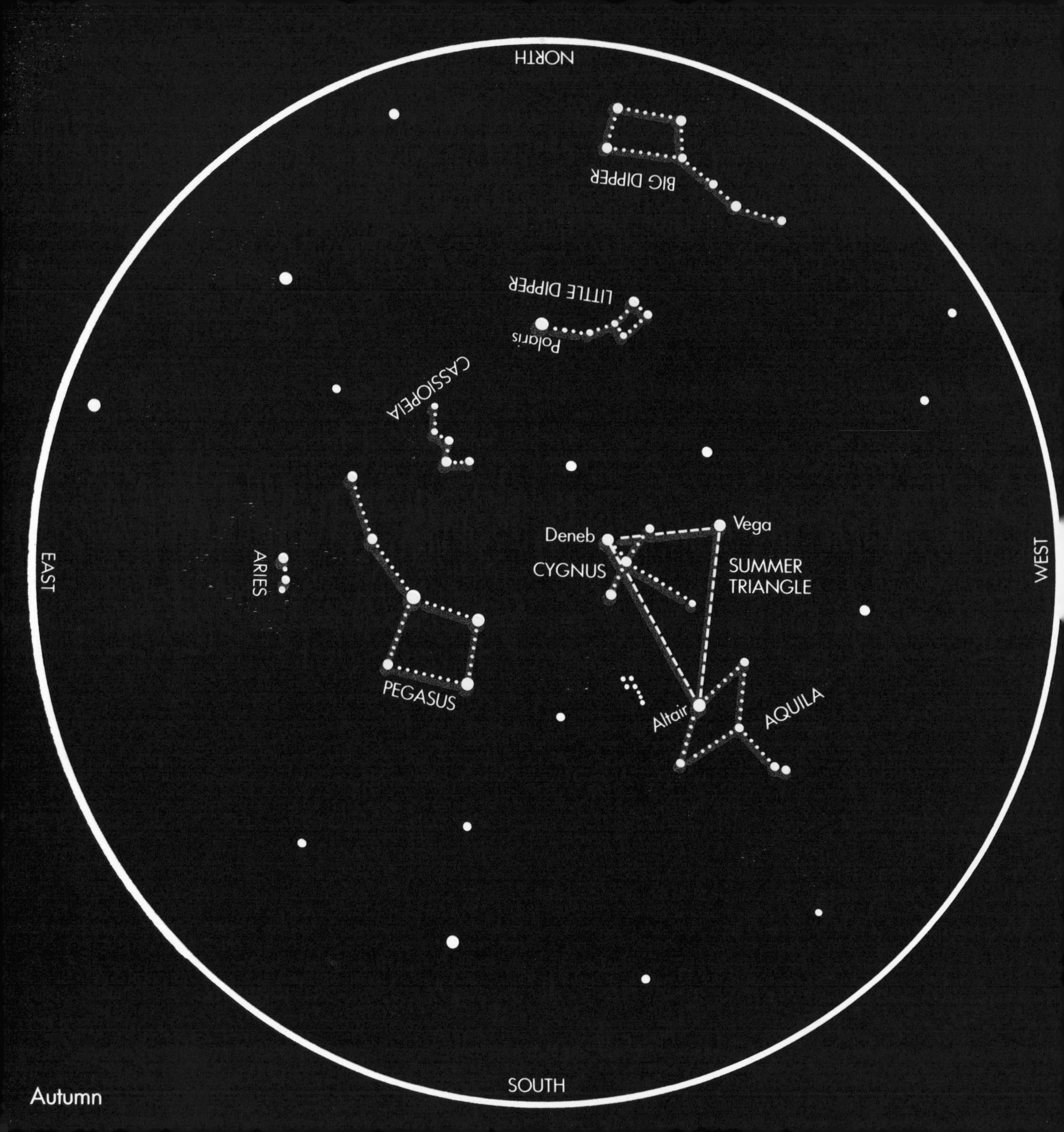
NORTH
BIG DIPPER
LITTLE DIPPER
Polaris
CASSIOPEIA
EAST
ARIES
PEGASUS
Deneb
CYGNUS
Vega
SUMMER TRIANGLE
Altair
AQUILA
WEST
SOUTH
Autumn

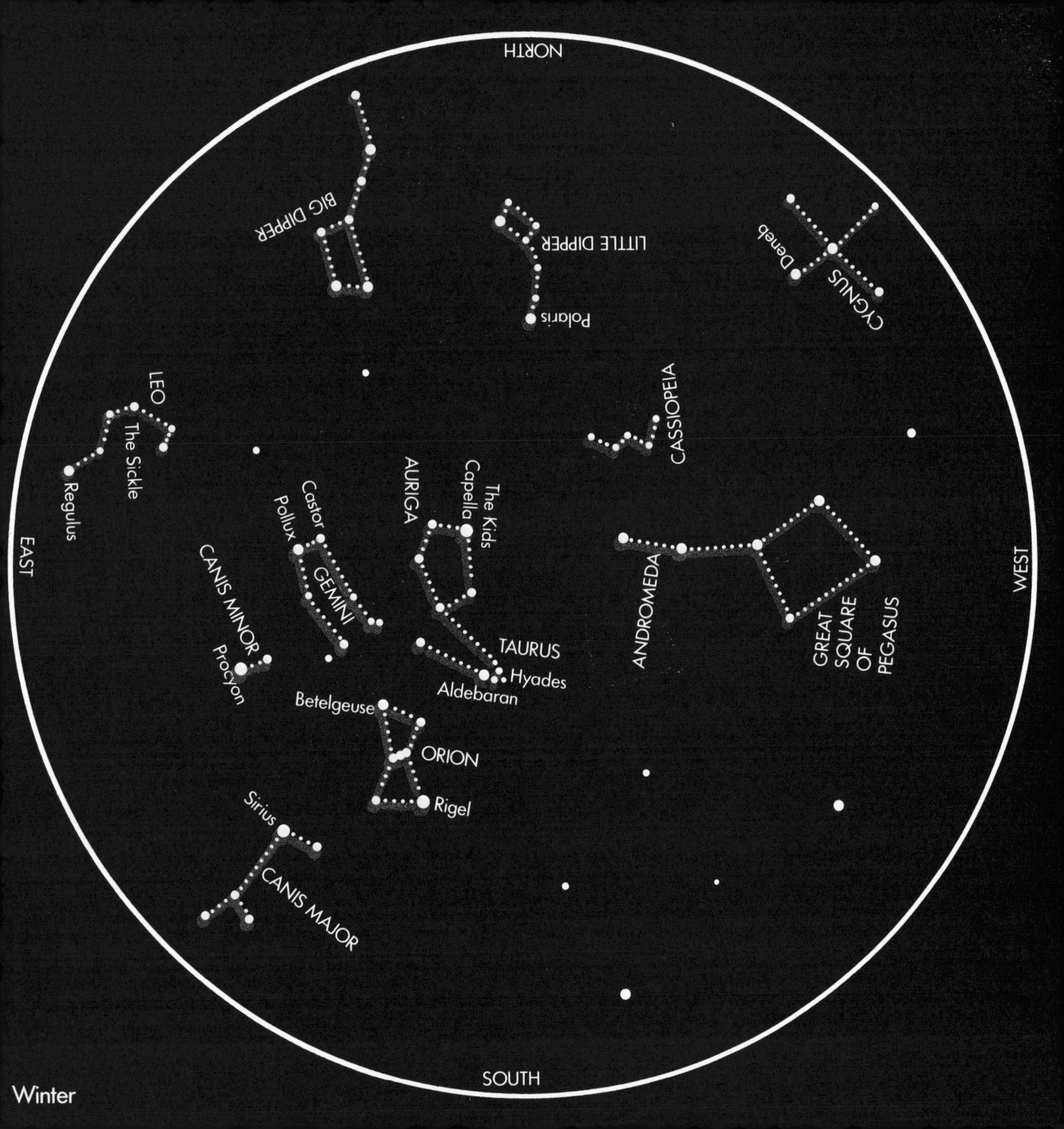
NORTH
BIG DIPPER
LITTLE DIPPER
Polaris
Deneb
CYGNUS
LEO
The Sickle
Regulus
CASSIOPEIA
AURIGA
Capella
The Kids
Castor
Pollux
GEMINI
CANIS MINOR
Procyon
TAURUS
Hyades
Aldebaran
ANDROMEDA
GREAT SQUARE OF PEGASUS
EAST
WEST
Betelgeuse
ORION
Rigel
Sirius
CANIS MAJOR
SOUTH

Winter